A Natural History *of* Sex

A Natural History *of* Sex

THE ECOLOGY AND EVOLUTION
OF MATING BEHAVIOR

Adrian Forsyth
Illustrated by Marta Scythes

FIREFLY BOOKS

Grateful acknowledgment is made to the following for excerpts appearing in this book:

Excerpt from *City Life* by Donald Barthelme. Copyright © 1969, 1970 by Donald Barthelme. Reprinted by permission of Farrar, Straus & Giroux, Inc.

Excerpt from *The Book of Imaginary Beings* by Jorge Luis Borges with Margarita Guerrero, translated by Norman Thomas di Giovanni in collaboration with the author. Copyright © 1969 by Jorge Luis Borges and Norman Thomas di Giovanni. Reprinted by permission of the publisher, E.P. Dutton, a division of New American Library.

Excerpt from "The Baculum." Copyright © 1975 by John M. Burns. Reprinted by permisson of the author from *BioGraffiti: A Natural Selection* by John M. Burns. Republished as a paperback by W.W. Norton & Company, Inc., 1981.

Excerpt from *Play It As It Lays* by Joan Didion. Copyright © 1970 by Joan Didion. Reprinted by permission of Farrar, Straus & Giroux, Inc.

Excerpt from *The Greek Myths* by Robert Graves. Copyright © 1958, 1981. Reprinted by permission of Curtis, Brown Associates, Ltd.

Excerpt from *Sex and Destiny: The Politics of Human Fertility* by Germaine Greer. Copyright © 1984 Germaine Greer. Reprinted by permission of Harper & Row Publishers, Inc.

Excerpt from *An Almanac for Moderns* by Donald Culross Peattie. Copyright © 1935, 1961 by Donald Culross Peattie. Reprinted by permission of Curtis Brown Associates, Ltd.

Excerpt from *Remembrance of Things Past*, Volume I, by Marcel Proust, translated by C.K. Scott Moncrieff. Copyright © 1934 and renewed 1962 by Random House, Inc. Reprinted by permission of Random House, Inc.

Published by Firefly Books Ltd. 2001

Originally published by Scribner's, New York, 1986
Published by Chapters Publishing Ltd., Shelburne, VT, 1993

U.S. Cataloging-in-Publication Data
(Library of Congress Standards)

Forsyth, Adrian.
A natural history of sex : The ecology and evolution of mating behavior / Adrian Forsyth – 1st ed.
[192] p. : ill. ; cm.
Includes index.
Summary : Essays exploring reproductive and sexual behavior in the natural world
ISBN 1-55209-481-2 (pbk.)
1. Sexual behavior in animals.
2. Sex (Biology). 3. Sex (Psychology).
4. Behavior evolution. I. Title.
591.5/ 6 21 2001 CIP

Published in the United States in 2001 by
Firefly Books (U.S.) Inc.
P.O. Box 1338, Ellicott Station
Buffalo, New York 14205

Produced by
Bookmakers Press Inc.
12 Pine Street
Kingston, Ontario K7K 1W1
(613) 549-4347
tcread@sympatico.ca

Design by
Janice McLean

The Publisher acknowledges the financial support of the Government of Canada through the Book Publishing Industry Development Program for its publishing activities.

Canadian Cataloguing in Publication Data

Forsyth, Adrian
A natural history of sex : the ecology and evolution of mating behavior

Includes index.
ISBN 1-55209-481-2

1. Sex (Biology). 2. Sex (Psychology).
3. Sexual behavior in animals.
I. Title.

QH471.F69 2001 571.8
C00-932904-8

Published in Canada in 2001 by
Firefly Books Ltd.
3680 Victoria Park Avenue
Willowdale, Ontario M2H 3K1

Printed and bound in Canada by
Friesens
Altona, Manitoba

Illustrations © 2001 Marta Scythes
Cover photograph: Wandering albatross (*Diomedea exulans*) pair in courtship
© Frans Lanting/Minden Pictures

CONTENTS

PREFACE

How to Look at Life

"Without speculation, there is no original observation."

Charles Darwin
in a letter to Alfred Russel Wallace

This is a book primarily for those interested in nature and secondarily for those interested in the biology of sex. It is specifically a series of essays about sex and courtship. More generally, it is an exercise in how to look at life, how to analyze and speculate on why something is as it is and not otherwise. This is also a book about the weird diversity of sexual behavior. It is an asking of why roosters crow and waggle their wattles, why a mite might mate with his mother and why the drone honeybee is designed to explode during his one and only mating.

It is this rich variation in sexual behavior that provides much of the beauty of life and determines the strength of evolutionary biology as a means of understanding it. Ecology and evolutionary biology depend on variation for the detection of patterns that become hypotheses, theories and even rules. We must speak of algae, bacteria, brain worms, parasitic barnacles, anything that seems to provide a clue or a question. For that reason, I have not shied away from our own species. Humanists, sociologists and those with a political dogma to foster often get upset when biologists ruminate on *Homo sapiens*. But I follow Thoreau: "I look on Man as but a fungus."

Humans are a special and complex species. Our sexual behaviors span the complete spectrum, from rigid, genetically based instinctive reactions to learned, culturally plastic behaviors. It may not be profitable to speculate long on the evolution of a polymorphous, culturally influenced phenomenon such as homosexuality, for example, but it seems reasonable to ask more concrete questions about such physical traits as the comparatively large breasts of human females.

Humans, however, are not the special interest of this book. The interest here is in the common principles that allow us to trace a relationship between the life of a fish on a coral reef and that of a jack-in-the-pulpit blooming in the shade of a northern deciduous forest. These principles explain the underlying attraction of the natural world for us, why we are not mute and indifferent to the world of other species, why our language is enriched with expressions like "proud as a peacock" that testify to a real, if intuitive, insight into the lives of other species.

Donald Culross Peattie, a naturalist and writer, once described the strong emotional impact of the first spring song of a frog: "On this chill uncertain spring day, toward twilight, I have heard the first frog quaver rise from the marsh. It is a sound that Pharaoh listened to as it rose from the Nile, and it blended, I suppose, with his discontents and longings, as it does with ours.... It speaks of the return of life, of animal life, to the earth. It tells of all that is most unutterable in evolution—the terrible continuity and fluidity of protoplasm, the irrepressible forces of reproduction—not mythical human love, but the cold batrachian jelly by which we humans are linked to things that creep and writhe and are blind yet breed and have being." Evolutionary biology is now uttering and seeking the forces linking us with everything else that has being. If we can discover the meaning in the trilling of a frog, perhaps we may understand why it is for us not merely noise but a song of poetry and emotion.

Unlike many other popular books on sexual behavior, this one is not concerned with how a behavior contributes to the survival of the species. A trait persists if it contributes to the reproductive success of the creatures that bear and propagate it. Thus we examine the self-interest and competition that result from natural selection acting on individual plants and animals.

This Hobbesian view of life may seem to be contradicted by sociality and cooperation in some species. Here is how evolutionary biologist Michael Ghiselin attacks this view: "The economy of Nature is competitive from beginning to end. Understand that economy and how it works, and the underlying reasons for social phenomena are manifest. They are the means by which one organism gains some advantage to the detriment of another. No hint of genuine charity ameliorates our vision of society once sentimentalism has been laid aside. What passes for cooperation turns out to be a mixture of opportunism and exploitation. The impulses that lead one animal to sacrifice himself for another turn out to have their ultimate rationale in gaining advantage over a third; and acts 'for the good' of one society turn out to be performed to the detriment of the rest. Where it is in his own interest, every organism may reasonably be expected to aid his fellows. Where he has no alternative, he submits to the yoke of communal servitude. Yet given a full chance to act in his own interest, nothing but expediency will restrain him from brutalizing, from maiming, from murdering—his brother, his mate, his parent or his child. Scratch an 'altruist,' and watch a 'hypocrite' bleed." We may be an exception to this bleak vision, but for the most part, this economy of nature is the guiding principle behind the logic of natural selection.

Bound up with self-interest is the theme of conflict of interest. Sexual behavior almost inevitably entails substantial conflicts of interest between males and males, females and females and males and females. The sexes differ markedly and fundamentally. A woman, for instance, may produce 400 eggs in a lifetime, but she may rear at most a few dozen children at an exhaustive physiological cost. A man can produce millions of sperm every day and sire, in theory at least, thousands of offspring at an exceedingly small physiological cost. A woman is almost always certain of her genetic relationship with her children; a male is never completely certain he is the father of a child. Such fundamental asymmetries in the costs and benefits of sexual behavior are responsible for the complex array of tactical and strategic relationships that characterize interaction within and between the sexes.

Many have wondered why males exist. By definition, a male is nothing more than an individual that produces small sex cells, the gametes used to make a new individual. No other masculine excrescences are necessary or sufficient to distinguish him from the female, the one that produces large gametes. The average male gamete, or sperm, is usually dwarfed by the female egg, and to some, it seems as if the male is a parasite on the female. Both sexes get the same genetic return from fertilization, but the material contribution of the male to the new offspring is a fraction of that of the female. This fundamental dimorphism in investment often extends into the realm of parental care, the cost of pregnancy and child-rearing being largely a female responsibility. Females bear the burden.

This is vexatious to those who believe that Nature is just. Fred Hapgood, in *Why Males Exist*, admitted that he "has tried to imagine ways in which the apparent paradox of the nonproductive male might be seen as making evolutionary sense." It was a work governed by his view that "when one looks at females in nature, one sees industry, progenitiveness and efficiency; when one looks at males, one sees the most amazingly elaborate forms of wastefulness. How is it possible that both genders have evolved together, as they must have? And are males really what they seem to be, an evolutionary frivolity, an extravagance with no practical point to them?" An earlier popular book on sexual reproduction, *The Mating Instinct* by Lorus and Marjorie Milne, concluded: "The male seems necessary. Not only the male but an abundance of them serving the species by increasing variation among offspring."

But why should males be "productive," and why need there be any "practical point to them"? Male gametes will evolve if they increase the reproductive success of the individuals that produce them. Nothing more or less is needed. As for the second point: Males and females are not necessary for sex. Sex and the genetic variation it produces do not depend on gender. There are many organisms, such as various algae, bacteria and fungi, that have sex without any dimorphism in gamete size. One cannot distinguish between males and females, merely individuals that are genetically compatible. There are even organisms that have sex with themselves. The paramecium reshuffles its genes within its own body to produce new genetic combinations. So there is no clear rule that says sex must have gender.

Most organisms, however, are recognizably male or female. An individual can make many small gametes (the male strategy) or fewer but larger gametes (the female strategy). They are two strategies that will be evolutionarily stable; that is, neither can outcompete the other. One produces either large gametes designed to fuse with a small one or tiny gametes designed to fuse with a large one. The female strategy produces large, competitive gametes with a high rate of survival and fertilization. The male strategy is based on producing as many sex cells as possible to increase the chances of finding a large female gamete. Any intermediate strategy does less well. Proving this is not a simple mathematical exercise, but such a proof exists. Those without mathematical acumen may take the argument on faith, bolstered by the knowledge that it is what the vast majority of species have opted for.

The important result of this is that an individual must invest in either a few large eggs or millions of sperm. Thus there must always be many times more sperm than there are eggs. Because of their redundancy, sperm must compete for access to those rare eggs. Most of what is masculine is determined by the quest for access to eggs.

In the first section of the book, the essays explore male strategies and tactics for increasing the number of offspring fathered. It is a story of adaptation and counteradaptation. In many cases, male-male and sperm competition will act to the detriment of the females that are the object of competition. In areas where biologists once analyzed behaviors according to how they might increase the strength of the pair bond, there is now a tendency to consider males and females as opponents. If a female can use the sperm from more than one male or gifts from male suitors, it may be in her best interest to mate with many males. This is clearly counter to the interests of each male seeking to monopolize her eggs. If she obtains sperm in an adequate quantity from a single male, she may be forced to fend off unwanted males that waste her time and may injure her. On the other hand, some females have evolved to exploit sperm competition and male-male conflict to increase their reproductive success. There is no reason to expect that what is good for the goose is good for the gander. In fact, the sexes differ so sharply in their strategies and tactics because of the egg-sperm dimorphism that we expect conflict to be persistently present between interacting males and females, just as we expect conflict between males and males and between females and females.

Our interest is in ultimate answers, that is, the evolutionary basis of a phenomenon rather than the details of its mechanism. Thus when we ask, "Why are ripe tomatoes red?" we can come up with two different kinds of answers. The proximate cause of the redness can be found out chemically and physically, and it can be said, "Ripe tomatoes are red because they contain carotenoid pigments [xanthophyll] that reflect red light." This does not address the ultimate cause—whatever makes red tomatoes selectively favored over tomatoes that are not red. To answer the question from this perspective, we ask how red affects the spread of the genes controlling redness, and we do so comparatively. We measure the effectiveness of red—as opposed to colors such as green or yellow—in attracting birds, which eat the pulp and disperse the seed to some sunny spot beyond the shade of the parental plant. Thus we are implicitly comparing alternative ways of attaining the ultimate goal of genetic representation in the future.

Some readers may wonder about the seemingly teleological and anthropomorphic linking of words such as "tactic" and "choice" to creatures with no more intelligence than a tomato. The wording employed here is in keeping with a venerable and, I believe, correct tradition in evolutionary biology. It saves us from the more exact but immensely tedious wording that goes: "Consider two genes in competition, one of which codes for red color, the other of which codes for green . . . ," and so on. To explain this wording to someone unfamiliar with its usage, I can do no better than to quote evolutionary biologist John Alcock: "The hypothesis that an animal has the ultimate goal of passing on its genes does not imply that the animal is consciously seeking to achieve that goal. Evolutionary entomologists sometimes write as though bees know who their relatives are, flies have strategies, and a bug can demand, 'like some errant macho Californian, proof of his fatherhood before paying out paternity benefits.' . . . Anthropomorphisms of this sort do not imply that insects have the cognitive powers of human beings. A male fly drawn to fresh dung surely is not aware that his chances of encountering receptive females are better there than at older cow droppings; nevertheless, this is the result of its preference. It is sufficient that the fly's nervous system operates in such a way that certain odors in certain concentrations are perceived and elicit certain responses."

Similarly, I have followed the widespread practice of applying words such as "rape" and "cuckoldry" to species other than ourselves. Some biologists object to these as anthropomorphisms and prefer less emotionally charged phrases such as "forced copulation" instead of "rape" and "kleptogamy" instead of "cuckoldry." Some see the use of terms such as rape to describe animal or plant behavior as a political act that buttresses the status quo and fosters sexism, arguing that "to name what flowers do as 'rape' is to specifically deny that rape is a sexual act of physical violence committed by men against women, an act embodying and enforcing the political power wielded by men over women." Not so. The forced copulation, or rape, of females by males, be they ducks or humans, is indeed a sexual act of physical violence. That it occurs and that biologists describe it as rape is in no way a justification for or an argument condoning it. Nature is full of horrible acts and events such as rape, murder, theft, starvation and disease, which human civilization has evolved to reduce and control. There is nothing to support the idea that what is natural is good.

The basis of biological determinism is the claim that "what was shall be" or even "what was should be." No serious student of evolution should ever subscribe to that belief, whatever his or her political persuasion. Humans have reason and will and can fly strongly against the adaptationist scenario. Some have happily thumbed their nose at the logic of the selfish gene and decided not to reproduce. Martin Daly and Margo Wilson, authors of an excellent biological book on sex, note: "Any species that includes substantial numbers of childless couples, exclusive homosexuals and religious celibates is sure to present some complications for the evolutionist!" But that does not stop them or us from speculating about broad patterns of human biology.

Karl Marx recognized the value of discovering what our natural behavior has been. He wrote: "The direct, natural, necessary relation of human creatures is the relation of man to woman. The nature of this relationship determines to what point man himself is to be considered a generic being, as mankind; the relation of man to woman is the most natural relation of human being to human being. By it is shown, therefore, to what point natural behavior has become human or to what point the human being has become his natural being, to what point his human nature has become his nature."

It should be admitted that there are dangers in trying to discover a Darwinian logic in every phenomenon. Darwin himself pointed out that it would be fruitless to look for the adaptive consequences of the color of fall foliage or the redness of blood. They are nonadaptive epiphenomena. We perceive them as striking, to be sure. Yet from the viewpoint of the organism's fitness, they are side effects of selection on other traits—the metabolism of chlorophyll or the oxygen-carrying efficiency of iron—but not selection on the colors per se. So we must keep in mind that explanations in evolutionary biology, however pat, are simply hypotheses to be tested and rejected in favor of a better idea.

Speculations about the adaptive value of a particular behavior or trait are sometimes decided as "just-so" stories that see adaptation and design by natural selection in every aspect of the world. Given enough assumptions, one can concoct a theory to explain any set of facts. This need not be a criticism if the assumptions and theory are open to testing and verification. So it does not bother us that so much of what is written about the sexual biology of our species or any other species remains highly speculative.

A comprehensive evolutionary ecology of sex is still a distant frontier. Observation and speculation both further the journey. We must follow the belief of Hans Zinsser, himself a serious scientist, who claimed that "scientific thought continually sets sail from ports of hypothesis and fiction." Bon voyage.

CHAPTER 1

Sperm Competition

"An inarticulate lucky stiff between
Paired spongy corpora casanova,
The baculum (or penis bone) of mammals
Lends firm support to a hard job."

John M. Burns
BioGraffiti: A Natural Selection

My friend Bruce collects junk. Like most field biologists, he has a collection of biological flotsam and debris, oddities he has picked up on his travels to remote places. The things ecologists gather usually look like the raw material for artwork by Henry Moore or Georgia O'Keeffe, and Bruce's tastes are no different. His Arctic collection consists of three bones. One of them is a human skull. Chewed by scavengers and coated with blue, green and gray lichens slowly mining its mineral riches, it is eloquent on the topics of nutrient recycling and existentialism. Another is a tribute to power, a sculpture of huge, flaring zygomatic arches, a steep sagittal crest and the massive chunks of bone that fuse to form a polar bear's skull. The last and oddest piece is a solid two-foot-long bone that looks like a policeman's club carved from ivory and bent slightly out of alignment by overly zealous application. It is the penis bone of a walrus.

People are usually surprised to hear that penis bones exist, since we are lacking in this regard. But odd and ostentatious as it may seem, the penis bone of the walrus is more than just a concrete form of male vanity. It is something that many males have had to evolve because of adaptations by the females and because of competition from other males. Male bats, shrews, moles, many carnivores—such as dogs, cats, bears, weasels, seals and sea lions—and most primates, including our

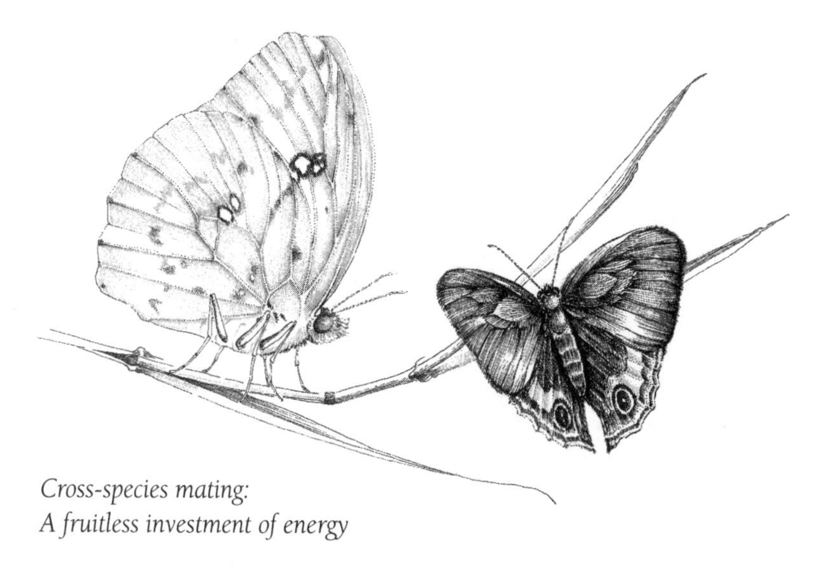

Cross-species mating:
A fruitless investment of energy

closest relatives, have them. Mammals such as humans that lack them are in the minority. The reason that so many males have had to evolve these structures is sperm competition for female eggs. The most plausible explanation for the evolution of penis bones, penises and internal fertilization is that they place the sperm closer to the eggs and make fertilization more likely.

From the female's viewpoint, internal fertilization gives her more control over who does the fertilizing and when. When a female frog or fish jettisons eggs into the water, any number of males can release clouds of sperm over them. With internal fertilization, females have the means to exclude the wrong sorts of males. Various kinds of insect, snail and other invertebrate females have long, complicated vaginas defended with flanges, flaps, sphincters, twists and curves that exclude males of the wrong species. This reduces the likelihood of a female mating with a male of the wrong species and wasting her eggs in producing hybrid and unviable offspring.

Not long ago, I watched a female sulphur butterfly flitting her way across the lawn looking for some alfalfa or clover on which to lay her eggs. This kind of butterfly is large, brilliant yellow and distinctive. Yet suddenly, she was assailed by a much smaller mottled brown ringlet butterfly, a male being grossly indiscriminate. He flew at the hapless sulphur female repeatedly, driving her down onto the grass, trying to copulate with her until she was finally able to escape. Such a union would have profited no one. The ringlet is in a completely different butterfly family, and the outcome of sulphur eggs mixed with ringlet sperm is sure to be a failure. But such a mistake is far more costly for the female sulphur—she has but a few eggs to dole out, while the ringlet male has sperm to spare and being overzealous costs him little. His mistakes are outweighed by the benefits of vigorous courtship.

As soon as females evolve mechanisms to put some distance between their eggs and the males that would fertilize them, a situation is created in which the male placing his sperm closest to the eggs has an advantage. Hence some males have evolved a penis with penis bone, muscles, ligaments and expandable pressurized sinuses to support it.

Possessing a penis is not without some disadvantages. Animals that have to swim or fly efficiently risk producing too much drag. Thus passerine birds, which have internal fertilization, do without a penis. They simply press their cloacal orifices together. Whales have to keep their penises tucked inside their body cavities when they are not in use; otherwise, they would create hydrodynamic drag in the water. Aquatic and aerial animals, which are constrained by friction more than terrestrial animals are, usually lack penis bones. The bones are entirely absent in whales, even though many whale species must maneuver tubular members well over six feet long. Bats, which also have long penises, usually have tiny bacula (penis bones), but some have lost them entirely. Human males, irrespective of how little clothing they wear, generally have to tie up their penises with string or bark or tuck them under loincloths or into sheaths made from gourds whenever they are on the move. A dangling appendage is at risk from thorns, snags and nettles. The risk of damage and transport problems are undoubtedly why most species that have penises have evolved erections which enlarge the penis only when it is needed for insemination or display.

A penis is necessary because the female genital tract is often a long obstacle course for the sperm that must traverse it. The length of the mammalian vagina is vast in comparison with the size of the sperm. Even when placed at the limits of the penis, the sperm may still have a long way to travel. Human sperm are only 0.06 millimeters long; that means it would take 500 of them laid end to end to total an inch in length. But the combined length of the female vagina, uterus and lower portions of the fallopian tubes, where most fertilization occurs, may be a dozen inches. Possession of a penis cuts this distance in half.

Nevertheless, a long journey remains for the sperm, and the male ejaculate has been designed to make the trip easier.

Males with sperm able to reach the eggs before the female mates with another male will enjoy a greater advantage. Males can influence the rate at which their sperm travel toward the eggs in several ways. The sperm may cooperate by forming into groups. Paired sperm are known in silverfish—insects that live around drains and eat the paste out of old books—and in opossums. In the opossum, which does everything by twos (the female has a paired vagina, the male a forked penis), the sperm, if separated, will swim endlessly in a circle until they run out of fuel. Together, like canoeists paddling faster by pairs, they may race upstream. In whirligig beetles, the rowing crew is more like a slave ship, with up to 100 sperm grouped together in a long, rodlike structure with flailing tails, rowing—wriggling, actually—along each side of the conglomeration. The curious feature of this is that only one individual sperm gets to fertilize an egg. Many of the rowers may be left out. Since the sperm often have different genetic compositions, we might expect selfish competition among the sperm of a male. But the competition from the more distantly related sperm of other males or the difficulty of getting there at all by oneself may favor cooperation.

Human sperm can swim at the rate of roughly an inch or less per hour, but somehow, they cover the distance from the base of the uterus up the fallopian tubes in much less time than their swimming rate would indicate. This is a distance that should take five to six hours, but sperm are present in the fallopian tubes within one to two hours. (Human fertilizations often take place not in the uterus but in the fallopian tubes beyond.) This is the result of some special properties designed into the sperm and the seminal fluid.

The sperm are aided in their journey by seminal fluids produced in the seminal vesicles, two saclike glands that lie behind the bladder and dump into the ejaculatory duct. Also contributing are the Cowper's glands, which add an alkaline secretion, and the prostate, which also adds an alkaline secretion. These alkaline fluids buffer the sperm against the acidity of the vagina and stimulate the sperm into activity. They make up the bulk of the fluid in the ejaculate. In some mammals, this may be a considerable volume. A boar may ejaculate 500 milliliters, roughly a pint, and his seminal vesicles may weigh half a pound each. This expense is the result of both male-male competition and the difficulty of navigating the long female reproductive tract. In the case of humans, few of the sperm ever make it to the egg. Half of them may swim up the wrong fallopian tube, and there is attrition along the way: Of the 300 million or so sperm in the ejaculate, only 2,000 are likely to contact the egg. And there the contest continues. The tip of the sperm contains an enzyme that must dissolve a way into the egg, and as soon as one sperm has penetrated, the egg changes its physiology and bars entry to all the others.

The seminal fluids ejected along with the sperm of mammals can influence fertilization by affecting the muscles of the female reproductive organs. Semen in humans and other animals is full of prostaglandins, hormonal compounds that serve many functions in different parts of the body. The rich concentrations of prostaglandins in male semen cause muscular contractions of the uterus that move the sperm toward the egg. This explains why sperm travel faster than they can swim. Prostaglandin-induced contractions must pump the sperm ahead. A high proportion of males who are infertile have a low concentration of prostaglandins in their semen.

The same tactic has also been reported in octopus and dogfish sharks. Their sperm come packaged in a spermatophore, a saclike bundle laced with serotonin, a powerful muscle stimulant that causes contractions of the uterus which, in turn, move the sperm.

Compounds such as prostaglandins in male semen may also influence the female's hormonal physiology to the advantage of the male. Scientists who labeled prostaglandins radioactively found that the prostaglandins readily cross through the wall of the female's uterus and enter the bloodstream. Studies are lacking, but it may be that these compounds influence the female's reproductive physiology. Studies of insects such as cockroaches and fruit flies have shown that substances in the male ejaculate stimulate the female to form eggs.

In many mammals, ovulation is induced by copulation. Some carnivores, such as weasels, bears and cats, and many insectivores, such as shrews and moles, have induced ovulation; that is, the female does not release eggs to be fertilized until she has mated. In species such as mink, ovulation is induced only by prolonged copulation. The mating of mink has been relatively well studied because of mink fur farming. Mink may copulate for the better part of the day, periods of activity interspersed with resting phases. During this process, the male may ejaculate several times. Indeed, the expression "goes like a mink" was and possibly still is used by some to describe a person with a strong sexual appetite. From the male's perspective, the function of multiple ejaculations may be to ensure that ovulation occurs shortly after he mates with the female. Multiple ejaculations are also described in various rodents and have been shown to influence the number of offspring in the litter that is sired by a given male. It may be that mechanical stimulation is enough to induce ovulation, but males may also have hormones in their semen that influence the female's reproductive physiology.

In species with prolonged copulation and induced ovulation, the female often mates with only one male if he is able to induce ovulation. Her estrus is short, and once her eggs are fertilized, she aggressively repels other males. A porcupine female, for example, comes into heat for only 12 hours. In mammals such as shrews, moles and many others, the female that has mated grows a membrane over her vagina, excluding entry of debris, disease organisms and unwanted male suitors. This places a premium on effective sperm placement and the prevention of access by other males to the same female.

One of the male adaptations designed to prevent other males from adding their sperm to the race is the mating plug. The mating plug is highly developed in insects, snakes, parasitic worms and certain mammals. We expect its development to correspond with the availability and dispersion of female mates. Male moles are an example of high investors in plugs. Female moles, like most burrowing mammals, are aggressively territorial. They chase other females out of their valuable system of runways and chambers, which provide food and shelter and are dug at great expense. This dispersion of females prevents a male from guarding more than one female at a time from the advances of other males. Accordingly, male moles have invested heavily in glands and secretions that produce a mating plug in the female after mating. The Cowper's and prostate glands in a male swell to a huge size during mating season, some 13 percent of his body weight, and their secretions mix, like epoxy, to form a gummy solid matrix wedged firmly in the female vagina.

Descriptions of ground squirrel mating plugs make them sound like the silicone rubber used as a seal around bathtubs and taps—a hard, rubbery mass that solidifies in the female and remains there for several days. Such devices are rarely fail-safe. Males have evolved penises capable of displacing these plugs. The penis bone may play a valuable role here, acting as a pry bar. In many smaller rodents such as mice, the penis bone and the penis exterior are curiously shaped with spines and fluting. These oddities have been extensively used by taxonomists, for example, to distinguish between the 19 different species of western chipmunks that look and act so much alike. For the rodents, it appears that these eccentric-looking penises are designed to dislodge mating plugs. In rats, it was found that males were able to dislodge some 69 percent of mating plugs. Males usually eat the plugs, which may indicate that they are nutritionally valuable and costly to produce.

Many other organisms simply package the ejaculate in a spermatophore that is hard to dislodge. Instead of having free-swimming spermatozoa, as we do, various organisms produce the sperm in a dense form that may be coated and mixed with gels, proteins and other molecules. Snails, for example, often produce a solid, intricately shaped spermatophore that matches the twists and turns of the female vagina and sperm receptacle and is coated with sharp backward-pointed spines which presumably hold it in place and prevent other males from supplanting it with their own. These may be formidably large—longer, when stretched out, than the length of the female's body. In the giant octopus, the spermatophore is a pencil-thick package more than a yard long. Insects, salamanders and many arthropods have taken the spermatophore route. Females have managed to turn this to their own good. Many have evolved digestive enzymes that dissolve the valuable proteins holding the spermatophore package together, and some just out and out gnaw them down directly. Thus they can convert the male sperm competition into a greater number of resources to invest in eggs and the rearing of offspring.

Some arthropods use a single gigantic sperm as a mating plug. Some insects have evolved enormous sperm. The common house centipede produces a sperm almost as long as itself, more than half an inch in length. In some featherwing beetles, sperm are twice as long as the adults and must be stored coiled up. When entomologists collect the female beetles, they can often see the tail of the sperm dangling out of the female's hind end. The giant sperm are often ornamented and have spines that may keep them lodged in place. This strategy necessarily entails a reduction in the number of sperm that can be produced. Some males have gone in the opposite direction and increased the quantity of sperm.

There is a clear relationship between how promiscuous animals are, how much sperm males can produce and the size of the testes relative to body weight. This correlation was first noted for primates. Chimpanzees, it was observed, have a highly promiscuous mating system. Females advertise their estrous state, and for four or five days each month, while they are in heat, they may copulate with many different males. Males, in turn, may copulate with several different females and often repeatedly on the same day. This sets up a situation of sperm competition. Males that can produce more sperm and mate more frequently will have an advantage over males with lower production. Accordingly, the chimpanzee has very large testes in proportion to body size. By contrast, the mighty silverback gorillas that defend a harem of many females have relatively tiny testicles. The dominant male gorilla prevents sperm competition by physically intimidating rivals and preventing them from ever having an opportunity to mate with the females in a harem. The reason for the correlation between testis size and mating system is clear: Males must economize on sperm and the sperm-production apparatus, matching supply with demand. It is often said that sperm are cheap, and indeed they are compared with female eggs; but the total cost of the ejaculate produced is not trivial. The male that economizes on his reproductive apparatus presumably has that many more resources available for building large, more competitive muscular and skeletal systems.

The most extreme practitioner of primate sperm competition may be the woolly spider monkey found in Brazil. I once had the opportunity to hear firsthand the story of their mating system directly from Katy Milton, an ecologist at Berkeley, the only person who has made long-term observations of woolly spider monkeys in the field. It is a story she tells with considerable gusto.

Woolly spider monkeys have a very loose social structure. Males and females may roam about alone or in small groups. The only stable association the monkeys seem to have is the one between mothers and their young. Milton could easily tell when a female was coming into estrus. As many as nine males would arrive in the female's foraging area and follow her, constantly copulating with her whenever possible. There is surprisingly little aggression among the males. Instead, they seem to save their energy and devote themselves to producing and introducing a large volume of sperm. The receptive female encourages this by actively soliciting copulations. It is known that she may mate as many as 11 times with four different males within 12 hours. Over her two-day period of receptivity, the number of times she mates must be proportionately higher.

This high rate of copulation is reflected in the design of the male genitalia. Milton believes the testes are larger than those of virtually any other primate relative to body size. As in humans, the penis lacks a baculum and is relatively large. Accordingly, the female becomes literally swamped with ejaculate. After a few copulations, the female has received so much sperm that the next male to copulate causes a gush of semen to be expelled from the female. Males are literally competing through sheer volume, displacing each other's sperm with their own. The exuded coagulate semen is high in protein, some 81 percent, and both males and females collect it from around the female's genitals and eat it, a behavior that makes sense, since protein is hard to come by for most rainforest monkeys.

The most glaring question is: What caused these monkeys to evolve the sort of mating system in which males are selected for sperm-producing ability, while other primates have males selected for their ability to fight and guard the females? In the case of the woolly spider monkey, the female has a great influence on the system. When she becomes receptive, she roams widely. Milton believes that this allows her to attract a large number of males. The males provide no parental care, so from the female's perspective, good health and genetic quality are the desired traits she must seek in the male. Since producing large volumes of protein-rich semen would be virtually impossible for a sick or low-quality male, the female that attracts large numbers of males and engages them in an insemination tournament can ensure that her eggs will be fertilized by a relatively fit individual. She lets male-male sperm competition do the work and expend the energy needed for mate selection and saves herself time and energy in the process. In doing so, she is the main architect of the male mating apparatus.

This correlation between testis size and mating system has also been documented for sandpipers. My friend Ralph Cartar, a Charadriiphile (sandpiper lover), recognized that sandpipers show almost every kind of mating system known for birds, from polyandry, in which one female mates with several males, to monogamy to polygyny, in which males mate with several females. By measuring specimens from many species, Cartar found that polygynous sandpipers have relatively larger testes than monogamous sandpipers. They copulate more often and need a greater sperm-production capacity.

Humans fall between chimpanzees and gorillas on the testis-size continuum. This suggests that we were once far less monogamous than we are at present and may have had a fluid, promiscuous mating system in which sperm competition played a substantial role. Either polygynous or promiscuous mating systems—systems that lead to a high rate of copulation—may favor large testes in males. Although many human cultures are polygynous, promiscuity may have been important in the past. Humans share some of the characteristics exhibited by other promiscuous primates in which many males may mate with a female and vice versa. Like the woolly spider monkey, we show reduced sexual dimorphism. Humans are among the least differentiated of all the apes. It is true that men are differently shaped than women because men do not give birth and that males have less fat and more upper-body strength, but males are not much heavier than females. And the difference in fat and upper-body strength may not be an adaptation for male-male combat as has been repeatedly suggested; it may simply reflect the fact that in virtually all preindustrial and agricultural cultures, males did the big-game hunting.

Humans lack the large, stabbing canine teeth that characterize species such as gorillas and other primates with polygynous systems in which males fight to maintain exclusive access to a female or set of females. This reverses an idea put forth by many writers on human nature who stress the polygynous character of human societies. Intense polygyny may be a relatively recent cultural development associated with an intermediate level of resource concentration that enables a few men to afford many wives and their children. High male parental investment would favor behavior that ensures the investment is in his own offspring, not in someone else's. Nevertheless, promiscuous mating systems have been described for preindustrial cultures.

The Inuit of the Arctic and the Hurons of the Great Lakes regions permitted a high level of promiscuity so long as it was reciprocal. Inuit wife exchange was common among friends and was considered a social-bonding device that cemented friendships and alliances. In Huron society, young women copulated with many different men, and when one became pregnant, she was considered to be the authority on who the father was. Elisabeth Tooker, an anthropologist who produced an ethnography of Huron customs, writes: "The girls vied with one another as to which should have the most lovers. If the mother found none for herself, she freely offered her daughter, and her daughter offered herself. The husband sometimes offered his wife, if she were willing, for some small present.... Often, a young woman might have 12 or 15 husbands, not including other men, for after nightfall, the young women and girls and the young men went about from one house to another, regardless of whether or not they were married." Such mating systems would cause intense sperm competition.

The rigid mate-guarding that has characterized Western and other highly materialistic societies with large inheritances and high male parental investment may be atypical and relatively novel. A survey of hundreds of human cultures reveals that female extramarital sex occurs in three-quarters of them and is common in more than half. In any case, humans have an extremely high capacity to produce sperm, one that seems out of alignment with female monogamy. It has been suggested that our high rate of sperm production may be responsible for the universally high rate of masturbation by human males, irrespective of their marital state. Masturbation seems like a wasteful and potentially maladaptive practice. Nevertheless, males of many primate species routinely masturbate. Bob Smith, an Arizonan entomologist and author of some highly creative work on sperm competition in water bugs and humans, points out that masturbation may be adaptive if sperm and seminal products have a shelf life. Or it may simply be that the human male's capacity for sperm production was forged in days when the mating system was more fluid than at present.

Sperm competition to this point has been discussed as the generator of larger or smaller testes in relation to the body size of the male. Logically, we might expect it to have generated absolutely huge testicles as well. If sperm competition were not a force, one would predict that the world's largest testicles would belong to the world's largest organism, the blue whale, which reaches lengths in excess of 100 feet and weights of 150 tons or more. But the blue whale's testicles weigh only about 200 pounds. The world's largest testicles belong to a much smaller whale, the right whale, which is about half the size of a blue whale. Right whales seem to have a promiscuous mating system in which several males try to copulate with a female simultaneously, rolling around her and hugging her for hours at a time. The result of this has been to produce testicles more than nine feet long—half-ton sperm factories.

CHAPTER 2

Penetrating Solutions:
Transvestites, Rapists and Dwarfs

"A man ejaculates some 350,000,000 sperms, and a bull about 1,000,000,000; of these multitudes, one may be used for fertilization."

Jack Cohen

Gamete Redundancy—Wastage or Selection?

Anyone who has traveled long on little money will have encountered a species that has taken sperm competition to extremes. It is a species that used to lurk in bat caves and in the lairs of large European mammals. Now, it finds life easier in the cracks and crevices of dingy *pensions* and fleabag accommodations everywhere. It is the bedbug, a flattened bloodsucker that creeps stealthily onto the skin of tired sleeping travelers, saps them and then leaves, as thanks, a blotchy, itching red memento. Wherever bedbugs dwell, they give the room a sordid odor that seems in keeping with their undeniably nasty reproductive habits.

The female bedbug possesses a seemingly normal genital tract. The male, however, is outfitted with an outlandish, formidable penis that he wields in an almost perverse manner. His instrument of insemination is a swordlike stabber that he drives into the female's abdomen, of all places. And although this seems off target, he nonetheless ejaculates there. The act is known, appropriately, as "traumatic copulation."

The sperm migrate into the female's bloodstream and swim or are pumped through her circulatory system until they lodge in a special storage gland. She holds them there until she has had a blood meal and produced a batch of eggs to fertilize. She seems to have adjusted to the male strategy by evolving the Organ of Berlese (named after the famous Italian entomologist), a special pad of tissue in the abdomen that assists in repairing the puncture wound. Berlese suggested that the act was of benefit to the female because the male injects copious quantities of sperm which the female may use as a nutrition source. At the time, and until recently, there was no evidence for this benefit nor an explanation for traumatic copulation. Entomologist Howard Evans remarks: "The image of a covey of bedbugs disporting themselves in this manner while waiting for a blood meal—copulating with either sex and at the same time nourishing each other with their semen—makes Sodom as pure as the Vatican. We must set aside Berlese's interesting theory as unproved, but having done so, we are left with no explanation of the advantage of traumatic insemination. Needless to say, it is not allowable to credit a mere bug with inventing all this simply for the sheer ecstasy of it."

It has now been shown that female insects can use male ejaculates for nutrition and for building their eggs. Butterflies, fruit flies and katydids can do this. Females have, at least in some cases, turned the harassment of sperm competition to their advantage by digesting and using a part of the male ejaculate as food. Indeed, in some species such as katydids, they adjust the number of eggs they allow a male to fertilize according to the size of the sperm package he provides as a meal for his mate.

Traumatic insemination is an evolutionary attempt to circumvent sperm competition through such mechanical devices as vaginal plugs and scoops (appendages that remove another's sperm from the female). For example, studies have shown that a male damselfly can remove from his mate almost all the sperm that was deposited by a previous male. Both plugging and scooping are circumvented by traumatic insemination, which bypasses the genital plugs and puts the sperm in the circulatory system, where it is unscoopable.

But what of the observation of homosexual stabbings and inseminations of male bedbugs? Is this merely indiscriminate male lust run amok? The African bedbug, *Xylocaris maculipennis*, does inseminate other males. But this homosexual rape serves conventional heterosexual ends. The sperm thus injected migrate to the victim's own vas deferens, or sperm duct. When the injected male mates with a female, he will use some of the rapist's sperm. One male forcing another to do his insemination chores is the ultimate in sperm competition.

Sperm transfer by proxy has also shown up in freshwater snails in the genus *Biomphalaria*. These snails are found in the Neotropics, and some are carriers of schistosomiasis, so their genetic makeup has been reasonably well studied. Experimenters have, for example, isolated genetic characteristics such as pigmentation and albinism that can be used to establish parentage in crosses. This has enabled them to show unambiguously that indirect sperm transfer can take place.

These snails are simultaneous hermaphrodites that normally exchange both sperm and eggs when mating. If an albino snail mates with another albino snail, we expect albino offspring. But when researchers crossed two albino snails, one of which had previously mated with a pigmented snail, some of the progeny of the albino-albino cross were pigmented. Apparently, sperm that an individual receives in its vagina can migrate through the snail's body into the male reproductive tract and remain viable, ready to move at the next mating.

When this was discovered, the investigators called it "sperm-sharing," a label that implies reciprocity, cooperation and mutualism. But given a chance, *Biomphalaria* cross-fertilizes, even though it can self-fertilize. This suggests selection for outbreeding—using foreign sperm—in which case, sperm-sharing would hardly be selected. Its sperm competition is probably akin to that of bedbugs.

Access is relative. For a male insemination strategy to evolve, it need only raise his success relative to that of his competitors. Or put another way, it can pay to lower spitefully the success of other males. We can call this "spite" because it costs both the victim and the aggressor something. It will be favored if it costs the aggressor less and increases his relative reproductive success. This may be the reason for the evolution of the transvestite-style behavior of *Afrocimex* bedbugs found on African bats. Males have evolved genital structures that mimic those of the female, and they stimulate other males to copulate with them. This may allow males either to eat the nutrients in the sperm they receive or to simply deplete the fertilization capacity of their rivals.

Selection to lower the mating rate of male competitors has led to another form of homosexual rape in which the victim receives a plugging. The acanthocephalan worm *Moniliformis dubius* is a parasite of vertebrate animals. While immature, larval acanthocephalans live in invertebrates such as cockroaches; they transfer to the gut or lung walls of such animals as lizards and rodents. Sperm competition in adult acanthocephalans is intense. One male may inseminate up to 17 females, and a single insemination may last a female for 104 days, which is roughly two-thirds of the lifespan of the adult male. Since the females all tend to mature at the same time, male-male competition is further intensified. Males lucky enough to mate with a female attempt to guarantee their paternity by cementing the female shut. They have a special set of glands that pack and cap the female's genitals.

Males will also cap and plug males. The experimenters who discovered this phenomenon explain it as a kind of spiteful homosexual rape. No sperm was found in the genital canal of males that had been plugged and capped by other males; the only sperm the rape victims had was stored in their sperm glands. This implies that the rapists did not simply mistake their victims for females, which they would have inseminated. Rather, it seems that they attempted to lower the reproductive success of their competitors by cementing them shut, a feat that would automatically raise their own relative success.

This kind of spite is also employed by some salamanders. Many salamanders court a female, and when she indicates that she is sexually receptive, they deposit a spermatophore on the ground for her to pick up with her genital opening. The female behavior is highly stereotypic and provides the male with reliable information about whether he should lay down his spermatophore. Sperm may be cheap, but a spermatophore of multitudes of packaged sperm can be costly. Unfortunately, salamanders are not perceptive creatures. This enables some males to act as spiteful transvestites. They imitate the behavior of a receptive female and lead a courting male on, inducing him finally to deposit a spermatophore on the ground, making him waste his resources and thereby raising their own prospects.

Chemically based female mimicry has been detected in the red-sided garter snake, *Thamnophis sirtalis parietalis*. In winter, the garter snakes den by the thousands in rocky crevices and caves in a few spots in Manitoba. When spring comes and they emerge, a frenzied mass mating takes place. An emerging female may be buried in a writhing serpentine spaghetti of a hundred males all trying to copulate with her. Only one male in a group succeeds. When researchers censused the mating masses, they found that in 14.5 percent of them, the object of male attention was not a female but a male. Studies of the pheromonal scent signals of the animals showed that the males being mobbed by other males were producing the same kind of scent as the females did. The males that smelled like females had fully functional male reproductive organs, so much so that when "she-males" were experimentally allowed to compete against normal males for access to females, they had a higher mating success. So attractive are these she-males that when one is present, regular males will ignore a true female and begin to court the she-male. This suggests a possible advantage for the she-males that chemically mimic females. In a mating ball, they may confuse and divert their male competitors, allowing the she-male individual to get close to the true female and mate with her. The high success of these chemical transvestites raises the question of why the trait does not spread until all males in the population exhibit it. One possibility is that there is a significant cost to the trait. She-males may be mobbed by males and waste much time when there is no female around. In fact, that is the case in the masses—14.5 percent of the total—in which only a she-male was found. Thus intense male-male competition creates a situation that favors the trait but also sets a limit on the value of the trait.

Transvestism can also be used for direct material gain. Randy Thornhill has shown that this is part of the behavioral repertoire of scorpionflies in the genus *Panorpa*. These large, elegant insects, which have netted, mottled wings and are common in the understory of rich deciduous woods, are not true flies, but they do resemble them. Males that have good hunting success capture other insects to use as nuptial gifts for the female, or they may secrete from their huge salivary glands a large gob of saliva, which females like to eat. Both of these gifts are used by the female to make eggs. She doles out copulation time in proportion to the size of the gift. The larger the food item and the longer she sits and feeds as the male copulates with her (always clinging to his bargaining chip), the more sperm he is able to transfer to her. Obtaining the food for nuptial gifts is not easy, and some males forgo the task. Instead, they assume the characteristic calling posture of a receptive female, and when a male flies up proffering a nuptial gift, the transvestite impostor seizes it and departs. Presumably, he then changes his guise and plays the part of the conventional male.

The third strategy of the male scorpionfly is rape. Some male scorpionflies, without ever bothering to collect a gift by either conventional or deceptive means, proceed to search for females. Thornhill observes that on finding a female, the male "darts toward her and grabs her with his large genital forceps. The male that succeeds in getting a grip on the struggling female then seeks to reposition her using his notal organ (a clamping device on the dorsum of his abdomen) to hold the female while he maneuvers into position to make genitalic contact. Although females appear to resist and often struggle free, occasionally a male is able to grasp the female's genitalia with his own and inseminate her. There is no food transferred from the male to the female in these cases, and the exercise has the appearance of forced copulation." In other words, rape has occurred.

Rape, or forced copulation, is well developed in some other groups—waterfowl in the Anatidae, the family of ducks, geese and swans, for example. Everyone has heard of how birds such as swans supposedly pair monogamously for life. But in fact, monogamously paired female waterfowl are often subjected to copulation attempts by males other than their mates. During these forced copulation attempts, females can be "exhausted, wounded or even killed." Such forced copulations may be perpetrated by unpaired males that have been unable to attract a mate and have little to lose and something to gain by forcing themselves on a female. They are limited by female resistance and attacks from her mate. It is also true that mated males will act as rapists.

In snow geese, which nest in tightly packed colonies, it has been shown that mated males will try to force copulation on neighboring females. They defend their own female mate until all her eggs are laid and then make attempts on nearby females, a mixed strategy that may result in their fertilizing only one more egg on average, but it is still a significant genetic payoff.

Access to eggs can be achieved by less overt methods than rape, gift-giving and bloody combat. There is a strategy that seems almost unbeatable, and I wonder why, ignominious though it may seem, more males have not pursued it. It is to become a resident dwarf.

Some males become miniaturized, living inside or on the females as a portable sperm bank. Darwin discovered this strategy when he worked on barnacle taxonomy. In some species, he found that the males had degenerated into tiny, almost parasitic forms consisting of little more than reproductive organs. These degenerate males led some observers to claim that the females were hermaphrodites. But actually, the male has evolved to become almost purely a copulatory organ permanently lodged in residence.

This kind of strategy has led to some of the most extreme forms of sexual dimorphism. In the vertebrates and the insects, extreme sexual dimorphism is normally a result of selection for the male fighting skills and weaponry used for acquiring a harem. When a male opts for an

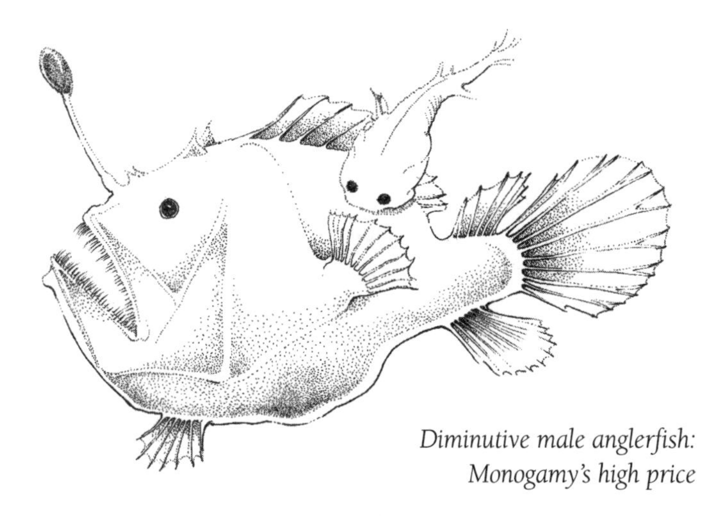

Diminutive male anglerfish:
Monogamy's high price

unshakable monogamy, the opposite dimorphism develops. One well-known case is the marine echiuroid worm *Bonellia*, common in the rocky bays of the Pacific Northwest and in the Mediterranean. When young *Bonellia* hatch, they are sexually undifferentiated free-swimming larvae. The transformation of their *tabula rasa* is effected by the bottom on which they settle. If it is unoccupied, the larva becomes a female. She grows into what is essentially a giant tongue almost three feet long with a walnut-sized body at one end. She uses her huge proboscis to pick up food from the bottom around her.

Males originate when a larva smells the scent of the female's tongue. If undifferentiated larvae are placed in an aquarium with chopped-up female tongue or an extract of the proboscis, they change into tiny males, scarcely longer than a twentieth of an inch. In nature, the dwarf male settles on a female proboscis, transforms, then lodges himself in or beside his mate's uterus. There he remains for life, leaking out sperm at appropriate moments.

This strategy is known in marine sowbugs and in some nematodes parasitic in rat bladders, but it is not found only in the lowly orders. There are dwarf male fish. Various species of deep-sea anglerfish are dramatically dimorphic. The female is a fearsome-looking predator of the abysmal dark lower reaches of the ocean. She has a thickset body and massive jaws well armed with teeth as well as a luminous lure, an organ developed from the first spine on her dorsal fin that projects out in front of her mouth and blinks, luring inquisitive victims.

Male anglerfish are so completely different that some were classified as separate species until taxonomists discovered them in connection with their female mates. The male starts off as an unremarkable surface-dwelling young fish. But after a while, he develops immense nostrils, and instead of a luminous lure, he grows a special tooth-bearing bone from his first dorsal spine. His mission is to find, perhaps smell out, a female and clamp securely onto her. In some cases, the attachment becomes permanent. The male literally becomes part of the female; he grows into her in such a way that their circulatory systems fuse and the male's mouthparts are completely useless for feeding. He depends entirely on the female. Presumably, he even reads the hormonal messages of her bloodstream to determine when he is to let loose his cloud of sperm.

The dwarf male has traded mobility and the opportunity for multiple mates for a relatively secure monogamy. This strategy has been explained by Michael Ghiselin as an adaptation to females that are rare and widely dispersed. This will be the case where the environment has a low energy content (the dark, deep sea) but where there is still fecundity selection on females to be large. Normally, the larger a female, the more eggs she can lay. Large size in a male, however, will not help him get more mates when they are rare and widespread. His mission is simply to find one and stick to her. Thus in the kinds of anglerfish and relatives common on coral reefs and in the echiuroid worms that feed by moving about and have a high rate of encounter, dwarf males do not exist.

The male, then, becomes a tiny resident not as a convenience to the female but simply because this has been the most successful male tactic. It is female dispersion and the ability of males to control the precious few large female gametes that make males into chest-thumping gorillas or degenerate dwarfs. For the bearer of superfluous and tiny gametes, the body is only a secondary concern. Access is all.

CHAPTER 3

Consuming Passions

"The ruling passion, be what it will,
The ruling passion conquers reason still."

Alexander Pope
Moral Essays

Passions often seem destructive to the individuals that fall under their rule. One summer, I observed a passion in bears that seemed to overstep the evolutionary bounds of good design. I had a few beehives within a couple of hundred yards of the house in a spot where there was frequent human activity. And there are plenty of hunters in the area, so the bears ought to be shy of humans. Yet the bear that discovered my hives was intent on plundering them regardless of the risk. My list of measures that do not dissuade a honey-lusting bear now includes placing a blaring radio on the hives, surrounding the hives with tripwires and clanging pots and pans, scattering the area with a pile of hair from a large, smelly dog and, most surprising, the formidable stings of the bees themselves. This is not anything exceptional. Leonardo da Vinci wrote: "It is said of the bear when it goes to the haunts of bees to take their honey, the bees having begun to sting him, he leaves the honey and rushes to revenge himself. And as he seeks to revenge himself on all those who sting him, he is revenged on none; in suchwise that his rage is turned to madness, and he flings himself on the ground, vainly exasperating by his hands and feet the foes against which he is defending himself."

Anyone who has inadvertently blundered into a large colony of bees or wasps might think the bear's honey lust is detrimental. Surely honey cannot justify the pain and the risk of anaphylactic shock and even blindness that 50,000 bees can inflict. Yet my bear returned each night to plague my hives until they were nothing but smashed frames.

This honey lust, like other consuming passions, may ultimately be adaptive. Concentrated calories are hard to come by in the wild, and in a bad year for foraging, a bear may wisely risk being stung for the opportunity to eat tens of pounds of concentrated calories and build up the fat essential for winter survival. It may be that the bears' determined pleasure-taking in the beeyard is not only an awesome desire but a genetically profitable behavior as well. It is one of biology's pleasures to try to understand the adaptive nature of seemingly excessive behaviors. I have in mind suicidal monogamy.

After the bear had pounded my hives to sticky splinters, I observed an even more dramatic display of self-destructive passion. Evicted along with the worker bees was a large number of male drones. The workers and queens regrouped, but the drones—that "work-shy gang," as Virgil called them—were helpless. Under normal conditions, they are indolent, ambling about the hives eating honey and doing little else. Now, with both honey and home gone, they buzzed haplessly around the garden flowers. I was in the garden with visitors when we came upon a male hovering, "droning" in midair, and on impulse, I snatched him up. I think I had it in mind to show my friends the size of his huge eyes. To my surprise and their amazement, he ejaculated instantly and expired in my hand before our eyes. At the moment of the drone's demise, a strange structure popped from his rear end. It had two yellowish inflated horns and a number of intricate flanges, curves and bristles. It was his mating structure, an exploding copulating device that my hand had inadvertently triggered.

Normally, the drone would have gone aloft, following the pheromonal scent of a hovering virgin queen until he joined a fray of other males, dozens, even hundreds of them jostling for access to the queen. The stakes are high for a chance to mate with the rare monarch, and when a drone is able to grab her, he wastes no time in fulfilling his true purpose. As soon as the queen opens her sting chamber to receive him, he explodes, his genitals bursting forth like a detonating grenade.

He explodes into the queen, and those horns and flanges wedge inside her. He also secretes a mucus that further clogs her. The male is thus paralyzed, his innards all askew. After a few seconds, he may separate from the queen with an audible pop and fall, eviscerated and dead, to hit the earth. Fulfilling his raison d'être has destroyed him.

On hearing of this procedure, one of my friends of course exclaimed, "What a way to go!" But spectacular as it is, this suicidal monogamy demands an explanation. How can natural selection have produced such a rigid, unbridled behavior that offers no second chances? Why should a male evolve without a modicum of self-restraint and concern for future mating opportunities?

The accounting is demanded not just of male honeybees. This hara-kiri style of mating is also well known in some spiders, which is supposedly how the black widow got her name. But the black widow, *Latrodectus mactans*, is not entirely deserving of her name. In many cases, she fails to eat her mate, and he may live to mate again. But there are other spiders that do seem to practice suicidal male monogamy by design in the way that honeybees do.

There is an orb weaver spider, *Araneus pallidus*, in which the size and mating antics of the male seem designed to place him directly on the fangs of the female. It appears virtually impossible for him to mate with her unless she bites him. Arachnologist Rainer Foelix describes the procedure used by the small male and the huge female. To set the scene, the female is hanging upside down vertically in her web, as orb weavers are wont to do, and the male approaches her with the goal of inserting his mating palp, which contains his sperm, into her orifice, known as the epigynum. Foelix writes, "He initiates copulation by jumping toward the ventral side of the female's abdomen and fixing one palp to the epigynum. During this process, he tumbles backward so that his abdomen rests right underneath the female's prosoma [upper body]. The female immediately seizes his abdomen with her chelicera [pincer] and, within a few minutes, starts feeding on him. Apparently, this is more a technical necessity than pure cannibalism. If the female is prevented from biting the male, he constantly slips off her abdomen and is unable to insert his palp."

The male of this species is so small that the female may not get much nutrition out of the process. However, she does eat him rather than just hold him in place, and he is forced into suicidal monogamy by the very design of his body and behavior.

The tendency of females to eat their mates is what has attracted most attention. Sexual cannibalism is a catchy phrase, but it is not clear how necessary or frequent a phenomenon it is. In some species, there is clearly a potential benefit to the female, and it is the standard mating procedure for some insect groups, such as various ceratopogonid midges. These are "no-see-um flies," those tiny specks whose bites leave a small but itching red spot on the human hide and whose presence has destroyed many an idyllic night's camping on an ocean beach or riverbank. Male ceratopogonids, like many flies, aggregate in mating swarms, hovering in a cloud over some light spot of ground or a treetop and setting up a visual and auditory signpost for females. These swarms are a favorite feeding spot for dragonflies and swallows. Even female ceratopogonids sally into these swarms to catch a male for dinner. One can believe Keats when he writes, "Then in a wailful choir, the small gnats mourn." For some species, female hunting is the standard mating technique. She flies into a swarm and drops down upon a husband. He is generally smaller than she is and feeble. She grips him face-to-face and stabs him through the forehead with her proboscis, then settles on a rock or other perch and injects him with a saliva laced with proteolytic digestive enzymes. Having given these time to do their work, she drinks up his liquefied interior.

Meanwhile, the dying male is going about his mission in life. His genitalia lock onto and into the female securely, so even as she casts away the empty husk of his body, his genitalia remain affixed to her. She is able to remove these eventually, but in many insect collections, one can see the male genitalia still attached to the collected, dried, pinned females of those species that engage in sexual cannibalism.

Common to these examples of the bees, the spiders and the flies is an excessive effort by males to guarantee their paternity by using their genitalia as a mating plug for excluding other males. In the spider, it is the palp that is used, and it is as elaborate as the apparatus of the drone honeybee. In species in which the female is capable of eating the male, the mating palp has an elaborate series of hydraulic bulbs and armatures and is detachable so that it, too, will lodge in the female as a barrier to other males.

Male readers may be gratified to hear that there are alternatives to losing one's vital parts in the pursuit of paternity certainty. Prolonged copulation is one means. Flies such as various bibionids are known for their long copulatory times. In the Northeast, they are called March flies, since they appear just as the snow thaws. In the Southeast, they are referred to as lovebugs, a less correct name but an inspired one derived from the fact that these flies can be seen copulating in huge numbers. Males earn the name by staying *in copula* for up to 56 hours, a feat made more impressive by the fact that as adults, they live for only two to five days.

Atelopus *copulation:*
Coitus noninterruptus

The function of this behavior has been studied in a Swedish seed bug, *Lygaeus equestris*, which sometimes remains *in copula* for 24 hours. Copulation is initiated and controlled by the male. His sexual appendages are such that the female cannot dislodge him unless he is willing. His apparatus consists of a long penis, approximately two-thirds his body length, equipped with hooks. Careful experiments have shown that the prolonged copulation is not used by the male to increase the amount of sperm transferred but to ensure that his sperm will not be displaced by another male's between the time he releases the female and when she begins to lay her eggs. Male copulation time increases as the number of males per female increases. In other words, when females are rare and males common, it pays to sequester the female. Also, the longer a male holds on to a female, the more likely she is to begin laying eggs soon after release.

Prolonged copulation reaches its apogee in *Atelopus* frogs, which remain clamped in intercourse for as long as six months. It does not matter that they are not actually copulating; the male is losing weight and wasting away, sacrificing his freedom to pursue other females in exchange for a better chance to fertilize the eggs of the female when she begins laying. (This must inconvenience the female also.) Prolonged copulation and exploding genitalia are all part of the same system of trade-offs. The benefit of a male achieving future matings is weighed against the costs and benefits of attempting to increase pater-

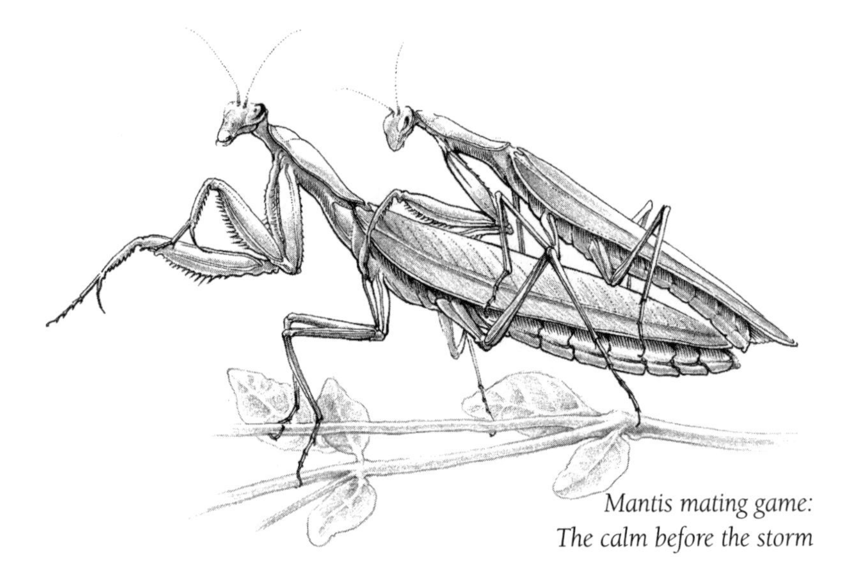

Mantis mating game:
The calm before the storm

nity certainty. What is unusual is that males should evolve willing death as part of their mating tactics.

Females can dislodge mating plugs, and males can displace other males. So why should a male risk putting all his sperm in one basket if cuckoldry is possible?

In the case of sexual cannibalism, it is often suggested that males have evolved to sacrifice themselves as a way of investing in their offspring. If the probability of obtaining a second mating is low, it might pay a male to feed himself to his mate, thereby increasing the number of eggs he will fertilize and perhaps the quality of his offspring.

This argument has been criticized on several counts. First, it is not clear in the cases of the praying mantis and the black widow that the males are the willing victims this theory would predict. A male black widow's weight is only 2 percent of the female's, and his contribution may not be a significant part of her total nutrient intake. A male mantis, however, is large and could be converted into a lot of eggs. Interestingly, in every case of sexual cannibalism reported, it is the sex which makes the heaviest parental investment that does the eating. Usually, this is the female. But there are males that have a consuming passion for their mates. Male polychaete annelid worms of the species *Neanthes caudata* assume the role of egg incubation and defense. Their investment is apparently high enough that females will fight over it. If a second female approaches the male while the first female is still present, combat may ensue. In this species, the male often eats the female after she has laid her eggs. Again, this suggests selection for the sex that invests more heavily to cannibalize the mate, but it does not indicate whether the eaten partner actively assists in the process. What is known about the incidence and behavior of mantises and most spiders under natural ecological conditions is too slight to evaluate the idea either way.

Steven J. Gould used this hiatus of evidence as the basis of one of his influential *Natural History* magazine columns. The disparity between the lack of evidence and the existence of theories and verbal arguments claiming sexual cannibalism as a male adaptation prompted his comments. It is worth quoting him at length, since he sets up a contrast between those who seek selective explanations of phenomena first and those, like Gould, who continually caution against the approach. Gould writes: "We have become overzealous about the power and range of selection by trying to attribute every significant form and behavior to its direct action.... Sexual cannibalism, with active male complicity, should have evolved in many groups (for the conditions of limited opportunity after mating and useful fodder are often met), but it has evolved rarely, if ever. Ask why we don't see it where it should occur; don't simply marvel about the wisdom of selection in a few possible cases. History often precludes useful opportunity; you cannot always get here from there."

Gould gives his reasoning for the argument earlier in the same article, where he points out that Darwin's proof of the agency of natural selection depended on the idea that history matters. Organisms are inevitably jury-rigged apparatuses full of historical artifacts and quirks. "Do not," he cautions, "rest your case on what might seem to be the most elegant expression of selection... for good design is the expectation of most evolutionary theories (and of creationism as well, for that matter).... Instead, look for the oddities and imperfections that occur only if selection based on the reproductive success of individuals—and not some other evolutionary mechanism—shapes the path of evolution."

Gould continues: "The largest class of such oddities includes those structures and habits which plainly compromise the good design of organisms (and the ultimate success of species) but clearly increase the reproductive powers of individuals bearing them." Finally, at the very end of the article, he argues that "our world is not an optimal place, fine-tuned by the omnipotent forces of selection. It is a quirky mass of imperfections, working well enough (often admirably), a jury-rigged set of adaptations built of curious parts made available by past histories in different contexts. Darwin, who was a keen student of history, not just a devotee of selection, understood this as the primary proof of evolution itself."

I believe Gould misrepresents the position of biologists whose work he cites in this column, those seeking a selective interpretation of sexual cannibalism. Biologists do not believe the world is an optimal place in Gould's sense. Otherwise, as one biologist put it, "butterflies would have machine guns"; organisms would be able to evolve anything of value to themselves. When a biologist speaks about optimizing a behavioral strategy or tactic, it means that trade-offs—costs and benefits to each option—exist. The outcome is, by definition, imperfect. For example, to achieve paternity certainty, a male must sacrifice opportunities for future mating. What could be a better example of Gould's suggestion to look for oddities that evolve not because they represent good overall design but because they enhance the reproductive success of the individuals possessing them? What could be a better example than the explosive suicidal genitalia of the drone honeybee?

To make selective sense of this phenomenon is not the same as saying that selection is omnipotent and finely tuned. Clearly, from the male's viewpoint, it would be better if he could both plug his mated queen and survive to mate again. The males of some bee species do have that option, including those of the closely related bumblebee.

The true task of the evolutionary behavioral ecologist is first to seek selective explanations for the diversity of behavior and then, if that fails, to seek explanations that have recourse to historical constraint, accidents, randomness and quirks of fate. The selective scenarios can be tested, but it is much harder to test and verify an argument based on accidents or unknowable history. In the case of the mating behavior of male bees, we do have a testable theory and predictions that are open to comparative analysis and experimentation. The gist of the argument is that there is a trade-off between the need to plug and the opportunity for second mating and that these depend on the operational sex ratio of the population. Honeybee populations contain thousands more males than queens. When a virgin queen takes over the colony, she inherits the hive and roughly half the worker force. Parental investment in males and females is roughly one to one in most species; thus in honeybees, vastly more males than queens are produced—the cost of a queen (the workers and hive) multiplied by the number of queens should equal the cost of producing a male multiplied by the number of males. Thus hives produce thousands of low-cost males for every high-cost queen.

Because of the relative scarcity of virgin queens, a male that mates has a low probability of ever mating again. Thus we can understand the sacrifice made to protect the one mating he does get. It is not our task to understand why he has not evolved a better way of doing it. It is our job to understand why, say, male bumblebees do not use this method. Male bumblebees can copulate repeatedly. In bumblebees, the queens disperse singly, they do not inherit any nest or workers, and their cost is not much greater than that of a male. Thus the numbers of males and females are not nearly so skewed as in honeybees. In other words, males have a high probability of achieving two or more matings. Accordingly, they invest more time in mate-finding and less time in mate-guarding.

Sexual cannibalism and suicidal monogamy have provoked much comment in scientific literature, and it is right that Gould should be irritated and moved to criticize the gap between the data on the topic and the adaptive explanations of the phenomenon that have been advanced. But selectionists are not simply marveling at the wisdom of selection and using sexual cannibalism and suicidal monogamy as examples. The selectionist turns to the bizarre and the extreme precisely because they offer the opportunity for the comparisons needed to test evolutionary arguments.

Suicidal monogamy is undeniably impressive. The behavior of male honeybees was known to the ancient Greeks, and they clearly found it striking. It is celebrated in the myth of the nymph-goddess of midsummer, Aphrodite Urania, who "destroyed the sacred king, who mated with her on a mountaintop, as a queen bee destroys a drone, by tearing out his sexual organs." That this procedure is so far from our method may explain why the Greeks thought the act must entail sacrifice on the part of the male and why ritual castration was conceived and used by the Phrygian priests as an expression of religious devotion and sacrifice to a higher value than self-interest.

I doubt whether we will ever make biological sense out of a religious passion so intense that it can lead even to self-castration. But there is cause to hope that the consuming and self-destructive passions of insects will submit to reason.

CHAPTER 4

Honest Salesmen

"They began little dances of suggestions and fear. These dances constitute an invitation of unmistakable import—an invitation which, if accepted, leads one down many muddy roads. I accepted. What was the alternative?"
Donald Barthelme
City Life

Redder than blood, almost exploding with iridescent light, they streaked through a patch of sunlight and then disappeared back into the gloom of the misted, mossy forest. They were male quetzals, the most beautiful birds in the Americas, birds so beautiful that the Mayan civilization decreed the death penalty for anybody who killed one. Then they were back, climbing sharply and suddenly nose-diving straight toward earth in a display that highlighted their streaming tails, almost a yard of resplendent emerald. There must have been a female nearby, but in the thick shrouds and folds of cloud-forest vegetation, I could not see her; she would lack the flaming crimson chest and dangling tail feathers of her suitors. It was spring, the hind end of the dry season in the mountains of Costa Rica, and the birds were back and courting. Males were everywhere, going through their exotic, absurd and beautiful song-and-dance routines.

On the emergent trees—dry stubs of now dead giants that rose above the other trees—three wattled bellbirds, arrayed in rich chestnut and white with strange black noodlelike projections dangling in front of their faces, were letting loose their metallic, clanging bong of a call. Females that settled beside a male were treated to a blast full in the ear, so loud they were literally blown off the limb. Lower down in the understory of the forest were the sounds of the male long-tailed manakin, a clear, flutelike *toe-le-doe* whistled all day long as the birds prepared their courtship dances. Let me describe the dance of the long-tailed manakins, for it epitomizes one of the most conspicuous yet mysterious problems in sexual selection.

The male has an arresting plumage, his chunky body dressed in basic black and highlighted with a bright turquoise-blue back, crimson crown and two tapering tail feathers that are as long as the body. The dancing court is a low vine or limb that runs horizontally a yard or two above the forest floor. This is the perch where two males will dance a duet, a pas de deux between a leading male and his accompanist performed for a female. There are variations in the dance, but usually, a successful performance follows this formula: The endless *toe-le-doe* calling, which the two males perform as a vocal duet, lures the female in. If the female lands directly on the perch, the two males turn to face her and begin, first one and then the other, to hop up and down, often jumping over each other leapfrog style. As they leap into the air, they flutter their wings, their tail feathers whip wildly and they utter a buzzing cry. The female herself becomes agitated, moving about, intently watching the males. The tempo picks up as the males continue to leap, flutter and buzz. They work themselves into a frenzy until, finally, the dominant male of the pair sounds a shrill note. He then slowly flies around the female, and the two mate. The performance is over.

This is all very entertaining but unexpected and hardly seems necessary. Proceedings such as these elicit a number of questions: What do all the wattles, tails, songs, dances and colors convey to the female? Why should she be impressed? Why should it matter to her? Are these displays just nonsensical ritual, or are they something that females can read and make use of? Would it not be quicker for her to simply mate and get on with the task of rearing her offspring? Biologists have been arguing about these questions ever since Darwin raised them more than a century ago.

Darwin pointed out that males might succeed in either of two ways: by conquering other males in battle or by charming the females. No one has ever doubted the importance of male combat, but the idea that males charm females has been more controversial. D'Arcy Thompson, author of the classic *Growth and Form*, ridiculed the idea because it suggested that there should be "vanity in one sex and wantonness in the other." Alfred Russel Wallace, the eminent naturalist and friend of Darwin, had a falling out with the great man over the issue. Wallace was troubled about the implications of the idea when applied to humans, but Darwin, unlike some of his contemporary followers, did not hesitate. Wallace objected: "A young man, when courting, brushes or curls his hair [and] has his mustache, beard or whiskers in perfect order, and no doubt his sweetheart admires them; but this does not prove she marries him on account of these ornaments, still less that hair, beard, whiskers and mustache were developed by the continued preference of the female sex." It is this issue that Wallace found galling, that female choice might have molded the secondary sexual characteristics of males. His alternative explanation was that males tended to be the gaudy, colorful sex because of inherently greater male "vigor" and a greater female need for concealment from predators.

Others objected to Darwin's ideas because they thought they implied that females were consciously evaluating male beauty and specific traits such as how well a mustache curled. But Darwin argued simply that females were attracted unconsciously to the overall appearance of some males more than others.

History has sided with Darwin. Most students of sexual selection accept that female choice exists. It can be demonstrated both with theoretical genetic models and in field conditions. In cases where the male is offering the female something of obvious material value, no one has any difficulty with the idea that female choice has influenced male evolution. If a male scorpionfly offers a female food she can use to make eggs, then she can accept or reject suitors according to the size of the nuptial gift. It makes straightforward adaptive sense. But what about the cases in which material resources are not an issue? Why should manakins dance? Why should male quetzals evolve tails a yard long? The tail of the quetzal is so long that when he is sitting on the nest inside a tree cavity, it protrudes conspicuously outside the hole. The reason for hole-nesting in the first place is usually to avoid predators, but the brilliant dangling tail of the male quetzal seems like a beacon for predators, a detriment to the male who must carry it and to the female whose nest may be robbed because of it.

It was these flamboyant and beautiful but seemingly deleterious characteristics that troubled Darwin and caused him to formulate his theory of sexual selection to explain them. The ability to attract mates by using such a trait might favor the evolution of the trait even if it lowered survival of the individual. What did the total number of offspring an individual produced matter? A trait would develop until the benefits it brought in mating success were opposed and balanced by the cost of reduced survivorship. Sexual selection could be seen as being in competition with natural selection for alternative forms of the trait. But why should females favor showy, deleterious traits that make males vulnerable to predators?

British geneticist and statistician R.A. Fisher was the first to formulate a model showing how female choice of a characteristic might result in a kind of runaway selection that leads to arbitrary and exaggerated traits which would lower a male's ability to forage and avoid predators and limit other important components of survival. Fisher said: "Imagine a male character that can be inherited—say, tail length—and has some initial value under natural selection. It might stabilize the bird under windy flight conditions, for instance. Imagine, then, an inherited female preference for this trait when choosing a mate so that they are more likely to choose males with a longer tail than those without. The sons from these matings between the preferring and the preferred type of female and male will then carry both the genes for the preference and the genes for the longer tail. Since the trait for a longer tail is favored by natural selection, the males that also carry the gene for female preference of long tails will have a greater survivorship. This means the inherited female preference will spread into the next generations. Eventually, the trait of female preference for longer-tailed males will become so common that any female which produces long-tailed sons will be favored, because her sons will get more mates than short-tailed sons. This means that the female preference, not natural selection for flying ability, is now driving the evolution of male tail length. If a number of genes are involved in controlling tail building and length, then the female preference can cause the process to run away, with tails getting longer and longer until countering natural selection stops the process."

Malte Andersson, a Swedish behavioral ecologist, has made a valiant attempt to determine whether this process actually occurs in nature. He experimented with the African widow bird, *Euplectes progne*, a species in which the strikingly marked black-and-red males grow tails up to four times their body length during mating season. Andersson cut some tails shorter and glued his cuttings onto the tails of other males to lengthen them; some males he left unbarbered. Then he scored their reproductive success. Males that had short tails or were

Male African widow bird: Runaway sexual selection or the straight story?

unaltered had the same mating success as before. But males with lengthened tails had increased mating success. This latter effect was not because the longer tails help the males in some task such as flying; long tails actually hinder male movement. Instead, it seems they succeed because of more effective and attractive courtship abilities.

Russ Lande, one of my friends from graduate-school days who now works in the stratosphere of evolutionary theory, has pondered this problem. In his abstract world, he finds it possible for female preferences to drive any kind of male characteristic, be it beneficial, neutral or deleterious in terms of natural selection. Lande, known to his friends as "Dr. Drift" (because of his interest in the way chance events may cause traits to "drift" in directions that are not favored by natural selection), has pointed out that small sample sizes and chance events such as mutation can establish a female preference for a male trait and cause it to become elaborated. In other words, the reasons why males are so flamboyant and why males of even closely related species often look and act so differently in their courtship rituals may simply be arbitrary genetic accidents that led to runaway sexual selection.

This finding makes many biologists uncomfortable. They would like to believe that females are not choosing arbitrary and meaningless characteristics, that females and other males can extract useful information from elaborate sexual traits, that the world is more intelligible than capricious. William D. Hamilton, another evolutionary biologist of acknowledged cosmic insight, has voiced his objection to the notion that male flamboyance is arbitrary: "I find it hard to believe that whole groups like pheasants and birds of paradise would have evolved and speciated successfully if their extraordinary sexual proceedings gain nothing for the participants except the plumes, wattles and supercoiled trachea, all of which seem so totally arbitrary, indeed, deleterious for coping with life in general."

The alternative to the idea of arbitrary runaway selection by female choice is the view that the displays and traits in question provide observers with accurate information about an individual's quality. The idea can be traced back to a notion known as the Handicap Principle, suggested by Israeli zoologist Amotz Zahavi. He argued that any male with a cumbersome handicap which survived to mate must be very fit and therefore a good type of mate to prefer. Unfortunately, models showed that a defective trait or handicap transmitted to offspring would not be favored by selection. However, the idea lives on in a modified form. Apparent handicaps can be favored precisely, indeed only, if they are costly.

A costly display is a signal that can't be faked. Therefore, males that are stronger than or superior to the average male can advertise the fact if they use a costly display which lesser males cannot afford. Both males and females, then, might be selected to reward flamboyant displays if they do distinguish high-quality males from the masses. In the case of the quetzals, for example, which live on a diet of fruit and an occasional insect or lizard, protein is probably a limited resource. A male that can afford to invest the protein needed to build a long, flowing tail is clearly healthy and perhaps genetically superior to a male with a shorter tail. It might also indicate that he has lived longer and been able to accumulate more resources. When these traits are genetically based and heritable, it would pay the female to evolve a preference for them.

There is considerable disagreement over whether there is any genetic variation in males that females could choose. Simple models show that if a genetic trait is favored by selection, the variation in the trait is soon winnowed away, and all males become the same. My feeling is that this and similar kinds of problems reflect the simplistic nature of the models and not the workings of nature. More complex models, including some by Russ Lande, show that a more complex world, one that allows for mutation, recombination and shifting environments, can replenish the stock of genetic variation, giving females scope for choice in male genetic quality.

In any case, females do choose some males over others. In lekking birds, such as the manakins, for example, where several males display in well-defined aggregations, females copulate selectively, preferring some males to a much greater extent than others. David MacDonald has shown that a careful analysis of the dance is a good predictor of how successful a male will be in attracting and mating with females. Males that call more attract more females; males that hover, a grueling test of strength, get more matings; and males with longer tails are older. Such correlations suggest that there may be scope for female choice and male-male competition to favor the elaboration of characteristics that separate the wheat from the chaff. The concept of honest advertisement, then, is an alternative to the notion of preference for randomly evolved traits, and it gives us a model to use in assessing whether any adaptive meaning lies in hitherto inexplicable courtships.

I think of the love darts of snails. At least 10 different families of land snails, which include hundreds of species, use love darts in courtship. Mate selection in these snails is already somewhat complicated because most land snails are simultaneous hermaphrodites; that is, an individual has both male and female reproductive apparatus. Normally, they reciprocate when they mate, each donating sperm and allowing their partner to fertilize some of their eggs. But before such proceedings take place, the prospective mates usually engage in an episode of mutual stabbings.

The love darts used are hard, sharp and up to half an inch in length, long in relation to the size of the snail and the foot where they are embedded. Many of the darts are complicated in design, and snail taxonomists use them to distinguish one species from another. Some are spiraled; others are multipointed. One of them looks much like a Middle Eastern ritual dagger, complete with curved blade and wrapped hilt. In some cases, they are used for multiple stabbings. In other cases, they are left embedded deep in the flesh of the recipient.

Most accounts dwell on the masochistic titillation of the act, but an evolutionist is hardly content with the notion of pleasure (or pain) for its own sake. I believe it is a form of honest salesmanship and female choice. When two hermaphrodites meet to mate, there is great potential for conflict and cheating. Each partner has both sperm and eggs to deal out. But sperm are far cheaper to produce than eggs, although they contribute equally to the new individual and are therefore equally valuable. Thus an individual might offer sperm but allow few eggs to be fertilized. Using this strategy would enable the cheater to mate with more individuals than would a snail that was free with its eggs. Being large and expensive, eggs are the limiting resource for land snails. By contrast, land snails' sperm are especially abundant, since they can be stored by the recipient for as long as four years. This means that a successful hermaphrodite ought to be unwilling to trade eggs for sperm unless it has some measure of reciprocity and some measure of the quality of the sperm donor. Enter the love dart.

Love darts consist of calcium carbonate—limestone. It is often a limiting resource for land snails, since they need it both to build their shells and to coat their eggs. Gathering the calcium carbonate and producing a love dart takes the snail a week or more. This means that possession of a love dart is probably a good indication of the snail's resources and ability to gather calcium from the environment. A snail deprived of its love dart can still mate, but this is not the issue. The important task of the hermaphrodite is to induce the partner to produce and to fertilize as many eggs as possible. What little is known about love darts indicates that they stimulate the female reproductive tract of the recipient. Stabbing may speed up or increase the amount of ovulation, especially if the recipient metabolizes the dart and converts it into egg production. This suggests that love darts are a masculine strategy, an honest advertisement that testifies to the male's suitability as a sperm donor and possibly acts as a source of male parental investment. Indeed, if a land snail is castrated, it still produces eggs and mates as a female, but it loses the ability to produce love darts.

Once one begins to think of courtship rituals and appearances as advertisements of quality, any number of new explanations becomes possible. William D. Hamilton and Marlene Zuk, in an article in *Science*, have used the logic of honest advertisement to explain why some birds have bright, expandable throat sacs, bare facial skin and long wattles and combs. They argue that it may be the way male birds proclaim their freedom from the many kinds of parasites and blood pathogens that infest birds. They reason that a poultry inspector is capable of quickly diagnosing dozens of poultry diseases by looking at the wattles, combs, feet and feathers of turkeys and chickens and argue that female fowl may have comparable acumen. They found that the more parasites a bird could potentially harbor, the greater was the development in healthy ones of ornaments such as wattles, which are richly supplied with blood vessels and brilliantly colored. Thus if parasitism is a potent threat, healthy males should evolve an appearance that advertises their health.

Hamilton and Zuk also found that the potential for harboring debilitating disease correlated well with the complexity of the male's song. It is now known that singing and other aspects of courtship demand great energy. Thus the male that sings long and loud proclaims his good health. Bob Montgomerie, a professor at Queen's University in Ontario, suggests that this is why the dawn chorus exists. After birds have spent the night shivering and burning up their stored energy to keep warm, they ought to go off and forage at first light. But instead, most territorial males engage in high-cost singing. Monkeys do likewise. By singing at the most costly, demanding time of the day, they are advertising their vigor. Hamilton and Zuk point out that if such advertising indicates good health, this alone, irrespective of any genetic aspect of male quality, would be something females ought to select in a mate.

Honest advertisement need not be directed only at females. Much of male courtship signaling is directed at other males. R.A. Fisher himself pointed out that "a sprightly bearing with fine feathers and triumphant song are quite as well adapted for war propaganda as for courtship." Male cricket songs, for instance, are usually thought of as signals that lure in females. But Chris Boake, in an article in *Science*, has shown that this is not always so. She studied the songs of Mayan crickets, *Amphiacusta maya*, in which groups of males sing together in hollow tree trunks. Males form dominance hierarchies that determine who has access to a receptive female. When Boake waxed the files of dominant male crickets to silence them, she found that, although they could still mate with females, they had to spend more time fighting off rival males. The dominant males still won their fights, but they wasted time and energy in the process and ended up with fewer copulations than they would have won by singing. In other words, it pays to advertise your strength to your rivals; otherwise, you will waste much in the process of affirming it.

The kinds of traits that ought to be used in this sort of advertising are ones which are hard to fake, such as the antlers of deer, which tie up as much energy and minerals as pregnancy and breast-feeding. Similarly, those to whom the signal is directed ought to have evolved a skepticism that prevents cheating. Usually, a status symbol is checked with some other behaviors. A deer with a large rack of antlers will still have to engage in antler wrestling, and if he is seen as weak, he will be subjected to escalated combat by his rivals and defeated regardless of the size of his antlers. That is undoubtedly why so many organisms do not simply evolve bright colors or long tails but display them in elaborate and costly energy-consuming performances.

The predictions, then, of the honest-advertisement view of sexually selected traits are rather different from those of runaway selection. One view says that the elements of courtship are packed with information about male quality, while the other says that the proceedings are arbitrary, the results of a genetic process influenced by chance.

In the end, I believe the two viewpoints will be amalgamated. I once read that in order for a musical composition or performance to captivate the audience, it had to be partly familiar and recognizable and partly novel and creative. That may be true of courtship and may explain the pleasure to be found in watching the antics of yet another bird species. We may find familiar adaptive explanations for why male manakins dance so wildly, but we may be unable to explain why different species of manakins have such radically different courtships. Such differences are the result of chance mutations that have been seized and shaped by natural and sexual selection. They are unique for every species. And so the special details of the long-tailed manakin dance, its choreography and colors, retain the captivating beauty of chance and novelty of forms that are impossible to anticipate.

CHAPTER 5

New Sneakers

"The race is not to the swift, nor the battle to the strong."

Ecclesiastes 9:10

Over a hill and down a dale from where I used to live is Lake Opinicon, a placid, shallow lake in eastern Ontario where I did most of my swimming. In its calm waters, I was often attacked by a most unlikely creature, the bluegill sunfish, *Lepomis macrochirus*. More than once when I was snorkeling peacefully by a colony of bluegills, one of them would come charging out, flaring his fins and gill covers, flashing his glowing orange chest and glaring at me in goggle-eyed outrage or even rushing in to knock me full in the faceplate.

Bluegills are not big fish—half-pounders are in the heavyweight class—so the contest was clearly unequal. And bluegills are the forage of pike, bass and otters, so their attacks on me always came as a surprise. The attackers are, predictably, territorial breeding males.

Bluegills nest in dense colonies in shallow bays all over eastern North America. Viewed from a boat or a canoe, their colonies look like a compact set of rings a foot or two across, all packed tightly rim to rim. Territorial males build these nests by sweeping out a bowl-shaped depression in the gravelly bottom. Females are attracted to the nests for spawning because the male owners provide egg-guarding services for the eggs they fertilize by chasing off predators such as bullhead catfish and by fanning the eggs to give them oxygenated water and some protection against egg-eating fungi. This is a task of constant vigilance, and it probably taxes a male considerably. Thus he is a jealous father, unwilling to risk being cuckolded or to allow his cherished eggs to be food for someone else's fitness. Lord of his ring, he challenges anything that comes too close, even a human snorkeler.

If you read a book about fish biology, this is the scenario you will get on bluegill breeding: Big territorial males build nests; females spawn in them. End of story. But a closer look at the Lake Opinicon bluegills has revealed that this is far from the whole story, a revelation that is part of a growing appreciation for the complexity of male mating tactics and strategies.

Territorial males first command one's attention. They are conspicuous. But in the case of the bluegills, there are usually more males than territories. Only 15 percent of the male adult bluegills hold territories. What about the other 85 percent? Are they simply nonreproductive losers? Studies by Canadian biologist Mart Gross and American biologist W.J. Dominey have shown that there are actually three kinds of breeding male bluegill: large territorial, or resident, males; a group known as "sneakers," which are at least one-third smaller and lack the bright orange coloration; and a third group of intermediate size known as "satellites," which seem to mimic females in both behavior and appearance.

Satellites, as their name suggests, hover above the colony. They wait until a female enters a nest to spawn and then sink slowly down between the female and the resident male. This confuses the resident male, who behaves as though he is still next to the female. The three circle around the nest, and when the female releases her eggs and the resident male jettisons a cloud of sperm, the satellite follows suit, adding his own genes to the mix. He then departs, leaving the resident male to provide the parental care. This is cuckoldry in classic form, a kind of genetic robbery in which the satellite steals the investment of the parental male. It enables the satellite to swim off, free to fertilize the eggs of other females.

Sneakers are less subtle. They simply dart into a nest when spawning is in progress, squirt out a jet of sperm and then depart. As with the satellites, their success depends on avoiding detection by the resident male. The resident, of course, is not a complete dolt. If he detects the cuckoldry ruse, he responds with a vigorous attack and damages the fins of the cuckolders by biting them. This is more than just a slap on the wrist; fungal infections are a great threat. One sees many a bluegill drifting around these lakes, barely alive, enveloped in a hideous, cloudy white shroud of *Saprolengnia* fungi. So the risk of being wounded and infected is not taken lightly. Nonetheless, sneakers and satellites enjoy considerable benefits from their thievery.

Sneakers and satellites invest no time and energy in growing large enough to fight for a territory and devote none of their resources to parental care, which enables them to become sexually mature at an early age—2 to 5 years—while resident males are 5 to 8 years old before they are large enough to hold a territory. Residents rely on large body size and strength to maintain their position. Sneakers and satellites are free to invest their resources not in muscle-building but in sperm. The cost of sperm is considerable.

As soon as sneakers and satellites reach sexual maturity, their growth rate drops by 30 percent. They develop testes that are 4 percent of their body weight, proportionally twice as heavy as those of territorial males. Since they are trying to flood, anonymously and quickly, as many eggs as possible with their milt, a small body with large testes makes sense. Nevertheless, being a sneaker does not seem like a glorious calling. It is far easier to admire the resident male, the doting, fearless father decked out in proud pigments.

Genetically, however, the resident male that has impressed so many biologists is no better off than the runty sneaker. Gross's analysis indicates that neither strategy—parental territorial or sneaker cuckolder—is able to outcompete the other genetically. They appear to exist in genetic equilibrium. Sneaking is, in the jargon of the day, an "evolutionarily stable strategy" in the sense that territoriality can never replace it. So long as territories are not hermetically sealed, sneakers can parasitize male parental care and propagate the genes that code for cuckolding behavior. This is a long way from the idea that nonterritorial males are simply losers, runts unable to make it. They are revealed instead as a truly and equally viable alternative to territoriality.

This discovery is a recent one, part of a larger trend in behavioral biology that seeks to quantify the behavior and success of individuals. Where ethologists in the past attempted to establish behavioral patterns typical of a whole species, behaviorists now admit that individual variation is highly significant and perhaps more interesting because it is the basis of natural selection and evolution, an attitude reflected in the heavy use of game theory by modern behavioral ecologists.

Game theory has been called "the science of conflict." It holds that there is no single "best" strategy in a given contest. An individual's success in following a certain strategy will depend on what strategies the other contestants employ. A good analogy is the child's game of paper-scissors-stone, in which the players flash their hands simultaneously, as a fist (stone), two fingers (scissors) or open (paper). Paper wraps and defeats stone, stone breaks scissors, and scissors cut paper. None of the three strategies is always best; each wins only in a certain context. The same principle applies to genetic and behavioral strategies.

When this perception is applied to previously studied animals, new observations and interpretations of their mating strategies result. Gelada baboons, *Theropithecus gelada*, for example, were first described in 1960 as having a social system organized as "one-male groups." These baboons live in troops of hundreds of individuals in the high grasslands and mountains of Ethiopia. Each troop is subdivided into groups of females that associate with one dominant male, or so it seems. The dominant male looks like a harem master. He is roughly twice the size of the female, sports a set of large canine teeth for fighting, is mantled with a thick, flowing mane and wears drooping whiskers that dangle eccentrically. It is what would be expected of an animal that must advertise his ability to endure and inflict throat-ripping bites. And fight he does, engaging in fierce battles to maintain or take over a harem. More recently, however, it has been revealed that some males never follow the macho route toward harem leadership. Instead, they establish a long-term relationship with one or two females. This might again seem like a second-best loser's option, but because these males don't engage in violent fights to take over or defend a harem, they live longer. Computer simulations suggest that the longer survival time means their submissive strategy results in a high lifetime reproductive success. As in the case of the bluegill, the two strategies appear to exist in stable equilibrium.

The evolution of alternative reproductive strategies has been favored by the high costs of conventional machismo and male weaponry. One of the better-known demonstrations of the cost of weaponry comes from the work of Valerius Geist on bighorn sheep. Bighorns are renowned for head-cracking duels in which two rams charge each other and smash their horns and heads together at full power. Success in these battles gives a male control over a group of females, but it is at a steep price. The large horns which males must build to withstand such collisions and which they use to signal their social status and fighting abilities require a large investment of mineral and energy resources. Geist measured the cost of this horniness by looking at sheep skulls he found in the field. Males with relatively large horns tended to die earlier. This might simply be because of the

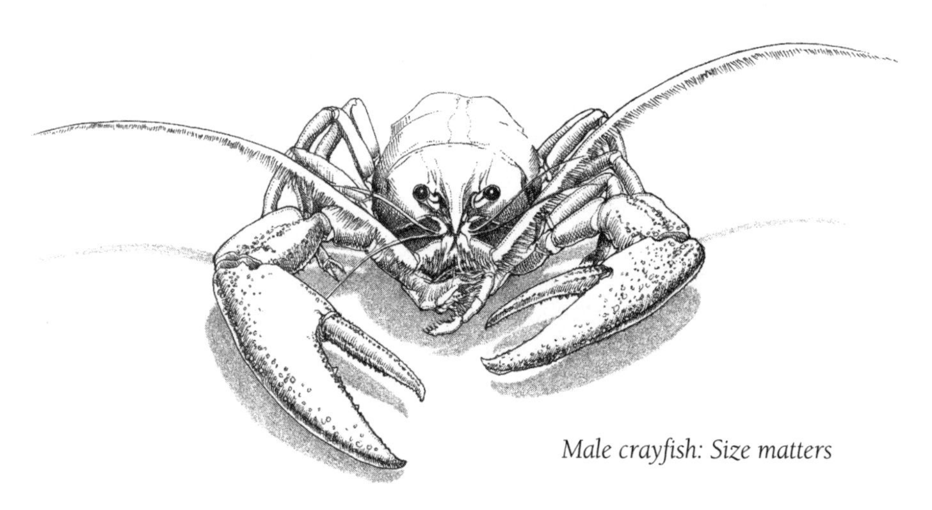

Male crayfish: Size matters

physiological price of the horns or, as Geist suggests, because the defensive behavior of a dominant male depletes the fat stores he needs to survive the cold of a mountain winter, thus raising his mortality rate to five to eight times that of his subordinate.

Not surprisingly, then, bighorn populations also contain sneaker males. They avoid fighting but attempt to rush into a group of females when a dominant male is distracted and copulate with a female on the fly. Or they may simply attempt to block a female from joining a harem and try to sequester her.

Another factor that sets a limit on the male arms race is predation. Male crayfish use their large claws in fights for females. On average, large-clawed males win these fights. But it seems that birds and other predators have the same feeding strategy as anyone who sets out to dine on lobster. The claw is a large chunk of meat that requires little handling time to process. Predation by birds is much higher on large-clawed males, a factor that must ultimately prevent the evolution of larger and larger male weapons by sexual selection.

Courtship signals such as flashing lights, dances, displays and songs are similar in intent to combat and weaponry. They increase access to females and intimidate other males. But they also attract predators. The love songs of male crickets are a familiar example. Most of us have heard males scraping their wing edges in a high-pitched ratcheting. It is the sound of summer. Most people also know that the hotter it gets, the faster the pulsing beat of the song. It makes crickets a fairly accurate thermometer, and it also points out that singing demands heat energy from the crickets. But the cost of this advertising includes more than calories. Predators use the cricket's song as a call to dinner.

Bill Cade, a professor at Brock University in Ontario, has described a system in which parasitic tachina flies home in on singing males and deposit their parasitic larvae on them. This explains, as Cade noticed, why some males are silent types. Silent males do not actively solicit females. Instead, they sit as satellites near a singing male and attempt to intercept any females that come crawling. This strategy is also recorded for a number of frog species and represents a remarkable convergence based on the fact that both groups use acoustic advertising and both attract predators because of it. The two different tactics—singing and silence—presumably fluctuate in success. Singing may be favored in years when predators and parasites are uncommon, and silence may be favored when predation and parasitism increase.

The novelty of such discoveries raises some questions. For instance, not only is the bluegill a common fish, but its reproductive biology has been the subject of dozens and dozens of studies, including some lengthy Ph.D. dissertations. Why did it take so long to see the sneakers? Theory determines what we see. Early ethologists were interested in comparing species and species-typical patterns of behavior, so individual difference, although interesting, was more of an annoyance that confused the attempt to describe a pattern common to all individuals of a population or species.

Another part of the delay can be ascribed to a pervasive misconception: that behavior acts to promote the survival of the species. Many early behaviorists misunderstood the principle of natural selection, and they interpreted behavior according to how it would benefit the entire species. Konrad Lorenz, one of the most famous ethologists, has, for example, written influential books suggesting that behaviors such as dominance and fighting evolve because they allow only the best individuals to reproduce, thereby benefiting and ensuring the survival of the species. This is simply wrong.

Natural selection is a straightforward principle that results from a few simple facts. Population sizes are limited. The members of a population produce more offspring than the environment can support. Some individuals vary genetically in their ability to survive and reproduce. Those individuals that are best able to propagate will come to make up future generations. In other words, individuals cannot act to promote the survival of the species except in a completely indirect manner. An individual must act in its own genetic self-interest. If a genetic trait or behavior is to persist through time, it must contribute to the reproductive success of the individual that carries it and transmits it to the next generation.

The two different perspectives lead to very different expectations. Let us return to the bluegills of Lake Opinicon. If individuals are behaving for the good of the species, we would expect only the largest and presumably the fittest males to get territories and reproduce. They ought to produce enough offspring to ensure the survival of the population and then stop. Or, to make it more elaborate, large males ought to compete for territories to ensure that the best contribute their genes to the future, and then reproduction should proceed peaceably. This is the sort of description inherent in the belief that only territorial males reproduce. Yet, in fact, this is not what occurs. Every adult male, large or small, tries to get more than his share of the action. The males fight, steal copulation, deceive one another, cannibalize eggs and manipulate each other. None of this makes any sense as far as the welfare of the species is concerned, but it makes much sense in the light of natural selection.

The tragedy of the good-of-the-species viewpoint is that it causes people to misread the behavior of plants and animals. The concept is tenacious. Hardly a day goes by that television doesn't carry some superbly filmed show on animal behavior whose meaning is obliterated by a commentator who explains it all as adaptation designed to promote the survival of the species. We can understand the persistence of this viewpoint. To act for the good of the species is an appealing notion. Our own species is more in need of such behavior than ever. But this invention is uniquely human. Nature's glory lies elsewhere, in individualism and all the diversity that attends it.

CHAPTER 6

Role Reversal

"The man in the street still tends to think of...entomologists as funny little creatures who caper over meadows with butterfly nets. When he finally realizes that biologists are working not only with the source of his daily bread but with his sex life, his ways of thought, his possible evolution, his most efficient extinction, then biologists will be even more hedged with throttling red tape than are modern physicists."

Edgar Anderson
Plants, Man and Life

Entomologists often feel irrelevant. The vast majority of insect species have no impact on the GNP or the price of bread. They do not cure cancer, they do not taste good, and many people think they don't even look good. All of this makes a public accounting difficult for the average entomologist. Typically, this accounting is demanded when he or she is working along some public road or byway. A car rolls up, a window rolls down, and from within, a head calls out, "What are you all doing?" The optimistic or naïve entomologist tries to explain, but a glazed look quickly comes over the face of the listening public. The explanation seems irrelevant to them, and they are soon motoring off, muttering about lunatics—and worse, if they suspect their taxes are paying for such arcane pursuits.

The remedy for this sort of entomological blues is to remember that the lives of insects may be philosophically profound, even radical. They are capable of changing what we perceive as the natural scheme of things. For many people, there can be nothing more natural than the notion that some types of behavior can be thought of as masculine or feminine—for example, the belief that males are naturally and inevitably aggressive and dominant and females are passive and coy.

This notion may have endured because Aristotle, one of the first authorities on comparative biology, argued for it. He claimed that in nature, "the male stands for effective and active and the female, considered as female, for the passive." Aristotle was not aware of the habits of giant water bugs and Mormon crickets.

Giant water bugs are disconcerting creatures to anyone who subscribes to Aristotle's view of sexual nature. And they are disconcerting to anyone averse to large insects. Some of them are larger than we like insects to be. South American ponds contain species the length of your hand and possessed of fearsomely long, strong raptorial forelegs and a powerful poisonous beak. These are the bugs that astounded writer Annie Dillard when she saw one suck a frog dry like a collapsing balloon down in Tinker Creek. But their eating habits are less fascinating than their sex roles. By Aristotle's account, they are reversed.

It is the female water bug that is active in courting, while the male is careful in his consortships. The male is concerned because of the female's desire to encumber him. She would coat him with eggs, neat rows of barrel-shaped canisters glued tightly across his back—his wing covers, that is. This entails considerable sacrifice on his part.

Giant water bugs are sit-and-wait predators. They lurk submerged and still in the water weeds and debris of ponds and streams, waiting for some hapless polliwog or newt to swim within reach. This habit is also protective. Underwater, they look like nothing so much as old brown, decaying leaves, a camouflage that must protect them from waterbirds and hunting mammals. They breathe air, but they linger at the surface for only a brief moment. Embryonic water bugs, however, thirst for oxygen. The eggs the female lays on the back of the male must be kept at the water's surface. The male must groom them and agitate the water around them to keep them well supplied with air, free from attack by fungi. Not only that, the eggs are heavy. He cannot float freely at the surface but requires the support of some stick or stone. And since his wing covers are plastered shut, the male, unlike the female, is rendered flightless by his pregnancy. This is of special concern to males that breed in temporary ponds and streams. A male thus burdened is more likely to be eaten and less likely to catch food or disperse to some better habitat than a female is.

A female can maximize her reproductive success by eating as much as she can to obtain the nutrients for egg production and by encumbering as many males as possible. Since one mating can provide all the sperm she will ever need to fertilize her eggs, she could save much time (useful for feeding) by simply laying her eggs as they are ready on whatever male she finds handy.

The male should have none of this. He is not merely a device for female convenience. If a male is to undertake the sacrifice—the investment in egg brooding—he must insist on paternity. His reproductive success will be maximized only by brooding eggs he has fathered.

Bob Smith, an entomologist and behavioral ecologist in Arizona, studied the courtship behavior of the water bug *Abedus herbertii*, a species common in the stock tanks and streams of the Sonoran Desert. He found it was the males that were careful and coy. A female would often initiate courtship, and the male would carefully insist on copulation before allowing her to encumber him. Indeed, he would allow her to lay only two or three eggs on his back before remating with her. These copulations are called "insurance copulations." Since a female may store the sperm from previous mates, a male repeatedly copulates to raise the probability that he will brood his own genes and not those of some other male.

Smith showed that this was a very real male concern by using some ingenious techniques. First, he found a genetic polymorphism—some individuals carried a gene that caused them to have a stripe down their backs, while others did not. This gave Smith a marker to establish whether an egg was fertilized by a striped or unstriped male. He then performed vasectomies on some males (this is, I believe, the only application of the technique to anything with six or more legs) and allowed them to interact with females. The females proceeded to mate with and burden the males that had had vasectomies using the sperm from a previous mating. In other words, females were capable of cuckolding the males.

The courtship rituals of these bugs fly against Aristotle's generalization. The behavior of the two sexes is tied not to whether they are male or female but to the relative costs of parental care. In this case, it is the male that has the most to lose from an ill-managed pregnancy. He faces the same sort of problem that confronts pregnant female mammals—metabolic costs increase while the ability to gather food and escape predators declines.

We could dismiss this as an aberration were it not for the fact that similar reversals occur in other organisms—in various birds, for example. The most famous of role-reversed birds is the American jacana,

Jacana spinosa, otherwise known as the Jesus bird for its ability to walk on water. (Actually, it spreads its great wide toes on floating water weeds for support.) It is a familiar sight on marshes and ponds throughout the American Tropics, an attractive bird with a rich chestnut-brown body and wings that expose a flash of lime-green or yellow when they are raised.

Jacanas are polyandrous. A female may mate with as many as four males. Each male broods a separate clutch of eggs. It is the female jacana that is the more pugnacious, and her territory encompasses the territories of several males. She is bigger than the male and uses her physical might to repel other females. It is even reported that a female jacana may oust another female, destroy her eggs and replace them with her own, an act equivalent to the infanticide practiced by male lions and monkeys when they take over a harem and kill the infants to bring the female back into estrus quickly.

Female jacanas fight to enlarge the size of their harems. Their reproductive success is limited by the availability of males to incubate their eggs. Marshes are extremely productive habitats with plenty of insect food, and it may be that a female is less limited by the availability of nutrients to make eggs than by finding attendants to guard and warm them.

Other birds can be monogamous but still show role reversal. Moorhens, *Gallinula chloropus*, form breeding pairs before they establish nesting territories. Females fight among themselves, especially if one female approaches a courting pair. These fights are reminiscent of rooster fights, in which the antagonists lean back and rake at each other with their clawed feet. Again, heavy females win most fights. But what are they fighting for? If they form pairs, there is a male for every female. The hens fight for fat males. The fatter the male, the better he will be at nest initiation and incubation.

Avian role reversals, also found in birds such as phalaropes, sandpipers and emus, were known to 19th-century naturalists. But the explanation of the phenomenon was lacking, and role-reversed species were viewed as bizarre exceptions. Darwin wrote of emus that "have a complete reversal not only of the parental and incubating instincts but of the usual moral qualities of the two sexes, the female being savage, quarrelsome and noisy, the male gentle and good."

Better still is entomologist Jean Henri Fabre's shocked account of the sexual roles of the cricket *Decticus albifrons*. "The male is underneath, lying flat on the sand and towered over by his powerful spouse, who, with her saber exposed, standing high on her hind legs, overwhelms him with her embrace.... In this posture, Master Decticus has nothing of the visitor about him. Have not the roles been reversed? She who is usually provoked is now the provoker, employing her rude caresses.... She has not yielded to him, she has thrust herself upon him, disturbingly, imperiously.... Master Decticus is on the ground, tumbled over on his back. Hoisted to the full height of her shanks, the other, holding her saber almost perpendicular, covers her prostrate mate from a distance. The two neutral extremities curve in a hook, seek each other and meet, and soon, from the male's convulsive loins is seen to issue, in painful labor, something monstrous and unheard of, as though the creature were expelling its entrails in a lump." That monstrous lump has proved to be a valuable experimental tool and source of insight. It was a male spermatophore that Fabre saw, a bundle of sperm and nutrients.

It is often said that sperm are cheap, which is true if a single sperm is compared with the cost of an egg. But sperm that come as a packaged deal with an ejaculate can, as we have observed, be an expensive gift. Spermatophores of the sort Master Decticus produced contain not only sperm but proteinaceous nutrients that females can eat or absorb. In various crickets and katydids, the female may bend around and eat most of the spermatophore; in butterflies, the female may absorb the nutrients through her internal organs. In either case, the female is acquiring male investment for use in her egg-manufacturing endeavors.

The cost of a spermatophore can be considerable. It may weigh half as much as the male's body. Think of that in human terms, and you will soon see that males ought to be selective about who receives this reward. In some butterflies with spermatophores, it appears that males discriminate against older females, whose worn, dull wings reveal them as a poor, high-risk venture in which to sink a large amount of capital. Male fruit flies, following a similar logic, prefer to court virgins because they will have more eggs to yield. Males do not have unlimited supplies of sperm. Many mammals, including goats, sheep, bulls, rodents and humans, as well as various fish, newts and insects, suffer sperm depletion. An ejaculation is followed by a period of reduced male fertility, and a recovery period is necessary for the sperm count to rebuild. In the case of various crickets and katydids, where the spermatophore is a huge proportion of male body weight, it is unlikely that males can ejaculate many times before they die. This sets up conditions conducive to role reversal.

Darryl Gwynne, a student of the mating systems of katydids and crickets, has shown that the spermatophore of the katydid *Requena verticalis* is used by the female to increase the size and number of the eggs she lays. Males, then, are a resource for females and can afford to be—indeed must be—choosy.

Gwynne has shown that in a related organism, the Mormon cricket, *Anabrus simplex*, in which males provide a large spermatophore, male choice exists. Not only do females fight when two approach a male, but when a male gets ready to mate, he weighs the female. Males seem to reject light females in favor of heavy ones.

This kind of male choice has also been observed in fish. The male three-spined stickleback, *Gasterosteus aculeatus*, and the male mottled sculpin, *Cottus bairdi*, if given a choice, prefer fat females over small ones. In each case, males invest heavily in parental care. The male investment acts as a limiting resource and explains why females behave in a "malelike" manner and fight for and solicit males. But there are other reasons for males to be choosy, even when they provide no parental care.

Males ought to be choosy about females whenever females vary in quality and males must invest heavily, not in parental care but simply in the task of acquiring a mate in the first place. This has been demonstrated for a snout weevil, *Brentus anchorago*, by Leslie Johnson and Steve Hubbell. These weevils perform one of the most spectacular mating tournaments in nature. I remember stumbling upon one in Costa Rica in the dry forest of Guanacaste. I came upon a recently fallen naked Indian tree, *Bursera simarubra*, a distinctive tree with smooth, coppery red-brown bark. The upper surface was alive with large, decorative weevils—long, thin black creatures striped with bright yellow. They were clustered in groups, some males using their long snouts to lever and bat other males away from the stout, lumbering females. Males were copulating whenever they could find the peace to do so, and some females were drilling their eggs into the wood. Since male weevils offer no parental care, it did not seem that males should be in any sense choosy. But Johnson and Hubbell found that they were.

Males of this species vary tremendously in size, as do females. Some adults may be more than 20 times heavier than others. In the large

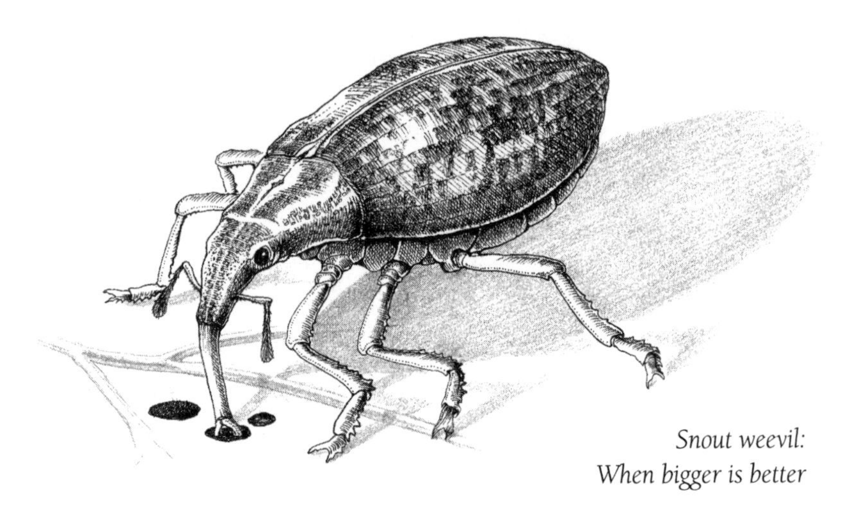

Snout weevil:
When bigger is better

male, much of the weight is due to his tremendously long snout, actually an elongated head, which is used for flipping and swatting rival males away from females. The largest males have a mating advantage, but it is not enough to give them free rein. They are not able or inclined to inseminate just any female indiscriminately. Given a choice, large males select large females, a choice that makes sense because large females seem to be more fecund or possibly because they may be resistant to sporozoan parasites that attack wood-boring beetles. Johnson and Hubbell made the remarkable observation that not only will large males fight selectively for females but they may even chase away small females. They noticed a large male swat a small, unguarded ovipositing female drilling her egg holes near a larger female that the male had mated. One might expect the male to trundle over and mate with any nearby available female. But the larvae of brentids (primitive weevils) compete, and if the actions of the small female posed a competitive threat to his larger mate, a male ought to treat her as a threat to his genes rather than as an opportunity. It is again a case where a male may be selected to improve the quality rather than the quantity of his offspring.

These studies are valuable in several ways. They do more than simply dispel the notion that there is such a thing as a masculine or feminine behavior pattern inherent to the genders. They also show that the sex roles played in courtship, territoriality and aggression depend on the relative investments of time and resources made by the two sexes. It is this theory, based on the costs and benefits of investments, that makes sense of masculine and feminine behaviors.

Many of these insights have come from the study of insects, and this is no accident. Insects are the richest group taxonomically and probably outweigh any other animal group in sheer biomass. It is to be expected that they will have the richest repertoire of roles. Some people may feel that they are, nonetheless, irrelevant. Humorist Will Cuppy once lampooned biologists by quipping, "The psychology of the orangutan has been thoroughly described from the observation of sea urchins." It sounds ridiculous, but that is the task of science—to find generalizations. Biological truths must transcend taxonomic barriers. Biology depends on comparisons of different organisms in order to create, test and refine its generalizations. It is a method that has proved successful in explaining the basis of gender-related behaviors such as who initiates courtship or who is territorial. Perhaps that is the relevance of the seemingly irrelevant entomologist, student of life most obscure. We define ourselves by what we know of other species.

In a scene in John Steinbeck's novel *Cannery Row*, Doc and Hazel are watching a patch of black stinkbugs waving their tails in the air. Doc, the biologist, comments to Hazel: "The remarkable thing isn't that they put their tails up in the air—the really incredibly remarkable thing is that we find it incredible. We can only use ourselves as a yardstick." The reverse is also true. To understand what it is to be human, we can use other species only as yardsticks.

CHAPTER 7

The Ecology of Abortion and Infanticide

"The women do not hesitate to use mechanical means or medicinal plants to cause abortion. Nevertheless, the natives feel, and show, very great affection toward their children."

Claude Lévi-Strauss
Tristes Tropiques

Mothers have a habit of murdering their children. Mothers of all kinds of organisms from cherry trees to humans abort embryos. They eat, strangle and abandon infants. What could be more anomalous? Destroying the carrier of one's genes ought to be selected against. Yet abortive behavior has evolved as part of the life history of many species.

Some students of abortion have viewed it as an epiphenomenal consequence of social stress or as an adaptation used to reduce population density. Biologist B.C. Wynne-Edwards popularized the idea that populations might evolve self-regulation mechanisms. This had intuitive appeal for anthropologists. Humans are creatures of some foresight, and no one likes to starve, which suggests a reason for humans to attempt to control their populations. Anthropologist Marvin Harris, for example, has explained infanticide as an adaptation human cultures use to optimize their density. Another example is J.B. Birdsell, an anthropologist who claimed: "Most demographers agree that the functional relationships between normal birth rate and other requirements (for example, the mobility of the female) favor abortion, lactation taboos, et cetera. The practices have the effect of homeostatically keeping the population size below the point at which diminishing returns from the local habitat would come into play." Nice as it sounds, evolution does not follow the same principles as resource management. Individuals may try to prevent other people from having children who would overburden the community's resources, but would they kill their own offspring for the common good? It can easily be seen that if abortive behavior in humans were genetic in origin, as it is in so many species, individuals who abort offspring for the good of the population would become scarcer in each generation than would those who selfishly reproduce at a high rate.

Abortion as population control does not make any biological sense for the individuals whose offspring are destroyed. Rarely, if ever, will individuals pursue strategies that benefit a population of weakly related individuals at great expense to the altruistic—in this case, abortive—individual.

Some anthropologists adopted a different view. C.S. Ford, an eminent student of human reproduction, wrote: "The widespread insistence upon not killing newborn infants suggests a tendency on the part of mothers to do away with their own babies. Were this not the case, it would be difficult to explain the very general development of sanctions directed against infanticide." But to posit that there is an instinctive desire to kill one's offspring makes no evolutionary sense. It is as far off the mark as the Freudian view that there is such a thing as a death instinct or a desire to commit incest with a parent. These constructs make no adaptive sense; they would be virtually impossible to evolve and embed as instincts in the human machine.

Both views—population-level selection for altruistic abortion and a pathological instinct to kill one's offspring—can be rejected as explanations for abortion. Instead, we must ask what the selective consequences are for an individual who does or does not employ abortion or infanticide. This approach can explain the practices not only of our own species but of those of sand sharks and milkweeds.

Far from being a pathological or altruistic oddity, abortion appears to be an effective female reproductive tactic that increases the reproductive success of those who practice it. All females and males begin life with far more gametes than they can ever transform into offspring. For a female, who pays a high physiological price for each offspring reared, there is a conflict between the quantity and the quality of the offspring. At some point, an increase in childbearing will conflict with the task of caring for a child.

The conflict between quality and quantity is well documented for our own species, especially in impoverished populations where resources are limited. UNICEF data show that a pregnancy improperly timed can lower the lifetime fitness of a woman. Pregnancies at either end of the reproductive lifespan, at too old or too young an age, are accompanied by substantially higher infant mortality. For example, children born to women under age 20 suffer twice the infant mortality of children born to women ages 25 to 29. If the interval between births is shortened, infant mortality goes up. A birth spacing of a year or less, for instance, produces twice the infant mortality of a two-to-three-year spacing. And as the number of children per woman increases, infant mortality increases. A child in a family of five siblings is twice as likely to die as one in a family of two.

In addition to the direct impact of excess childbearing on infant survivorship and health, childbearing physically drains the mother. In a species such as ours, in which parental care extends for many years, it is important that the mother survive long enough to complete the task of child-rearing. In other words, producing more children may actually lead to fewer grandchildren, great-grandchildren, and so on. Thus it is possible for a female to increase the number of children she bears but still leave fewer descendants than a female who opts for a strategy of fewer but fitter offspring.

Abortion and infanticide are influenced by resource availability. This has been neatly demonstrated with tomato plants. If a plant is shaded or has leaves removed, then fewer resources can be gathered. The plant's response to this treatment is to abort some of its flowers and fruit. The aborted progeny cannot be explained as genetic defects, because when they are gathered up and placed in a nutrient growing medium, many of them become large, healthy tomato plants.

In mammals, it is a common reproductive strategy to reabsorb embryos if the mother's health or feeding success declines. Lagomorphs, such as rabbits, and rodents such as various mice, rats, voles and beavers possess the ability to terminate fertilized embryos physiologically and absorb them back into the body. A rat may become pregnant while she is still nursing offspring, but if her milk demand suddenly rises, she will reabsorb the implanted embryos.

Infanticide is a less efficient method of cutting future investment costs, yet it, too, is widely practiced by rodents. The size of mouse litters varies. Where infanticide and cannibalism occur, litter size is usually above average and the killing tends to "normalize" the number of offspring, reducing it to near average. Experiments manipulating the growth rate of infants, their number and the lactation abilities of the mother show that a mouse with eight babies may gain by killing one infant and distributing her milk among the remaining seven.

The difficulty of producing enough milk for more than one infant is often cited as a reason for twin infanticide in hunter-gatherer cultures; when nomadism is involved, simply the problem of transport has been the justification. It is easy to forget how stringent these demands have been for most of our history, but if hunter-gatherers are any model, they are stringent indeed. Estimates of the infanticide rate for nomadic Australian Aborigines run at 20 to 40 percent.

One interesting feature of human infanticide is that it is usually prejudiced against females. In Inuit cultures, sex ratios may be so distorted that there are two males reared for every female. In the upper classes of India and China, sometimes virtually every female infant was destroyed. Selective female infanticide is also reported for native cultures in South America, New Guinea and Australia. This is not what one might expect on the basis of a superficial glance at their mating systems; most are polygynous, and high-status males have more than one wife. This means there is a surplus of breeding males. From the viewpoint of population efficiency, these males ought to be killed off. To some extent, they are: Warfare, fighting and hunting accidents give males a much higher adult mortality than females in these cultures. But even so, infanticide is selectively female.

Female infanticide seems to be an economic strategy employed because it affects the resources of the parents. In India and China, the practice was encouraged by the large dowries needed to marry off daughters. Rearing a son meant that the family would receive a dowry instead of having to pay one out. In more "primitive" cultures, traditional female labor either had no monetary value or was less important than the resources and success of male tasks such as hunting and working for pay. And in a patriarchal system, a family composed mostly of sons would be stronger than one composed mostly of daughters to be married off and dispersed among many lineages.

Resources, however, are not the sole factor controlling the frequency of infanticide and abortion. Quality control of offspring may also necessitate abortion and infanticide. Again, some anthropologists suggest that elimination of defectives was a way to keep the population pure. However, it is more likely that when resources are limited, parents prefer to invest in healthy rather than defective offspring. Human cultures, particularly the so-called primitive ones, do remove most children born with obvious congenital defects, and it is the parents who make the decision.

One of the consequences of sexual recombination, in which the chromosomes of different individuals are mixed, is that some of the new combinations are defective. So common are these defects that the human uterus seems designed to filter out a large number of offspring which do not make the grade. This seems to be the only explanation for the fact that even in well-nourished, healthy women, natural abortions are common. It is estimated that at least 25 percent and as many as 75 percent of all fertilized embryos fail to come to term. Regardless of their moral, ethical and cultural beliefs, women routinely have spontaneous abortions because their bodies physiologically reject the fertilized embryos. This makes sense, considering that the average woman has 400 eggs to choose from and millions of sperm but the resources to rear only a handful of children. The uterus ought to be choosy. The fact that spontaneously aborted embryos contain a high proportion of genetic abnormalities suggests that genetic screening by the uterus is the explanation for the high rate of natural abortion. It has been suggested that the rise in the probability of having a Down's syndrome baby as the mother ages is a direct result of the uterus losing its ability to screen out defectives.

Genetic screening may not only increase the efficiency with which maternal resources are invested but also act as a form of sexual selection. This will be especially important for females denied conventional forms of mate choice. Females ought to be relatively selective about whom they share their gametes and resources with. An animal may be able to assess directly the quality of a mate or at least allow a process such as male-male competition to do the winnowing. For plants and possibly for sessile organisms such as oysters, however, mate selection is difficult. Females must make do with what is borne on the waves or on the breeze or on the bellies of bees. A female plant may get some choice from the male-male competition between pollen grains, which must bore their way down through the pistil to reach the eggs, an act that may be a test of their vigor. But growth of a pollen tube may be a poor measure of how good a combination the male and female gametes will make. The female solution seems to be to fertilize extra ovules and do the culling later.

A cherry tree has this built into the female physiology. Every cherry fruit always starts out with two ovules. Both are fertilized, but only one survives to become the seed. The other is starved out of existence as the seed matures. Other trees take this to great extremes. Kapok trees, *Ceiba pentandra*, of the Tropics, whose fluffy dispersal devices serve as stuffing for pillows and life preservers, abort on the order of 1,000 flowers for every fruit they bring to maturity. In highly outbred plants, where many new and different genetic combinations are produced, abortion is high. In habitually inbred self-compatible plants, pollination involves little genetic variety; there is less variation to choose from and correspondingly little abortion.

Evidence for an interaction between genetic quality of offspring and maternal resources also comes from agricultural plants. Cucumbers, for example, can be either self-pollinated or cross-pollinated. Vegetable growers or breeders who wish to obtain self-pollinated seeds must pick off all the outcrossed cucumbers. If they do not, the cucumber plant will shunt all its investment into the outcrossed cucumbers and abort the self-pollinated and highly inbred fruits. This makes sense for the cucumber. Normally, inbred cucumber seeds are less viable and vigorous than the genetically diverse outbred seed crop. The same is true of macadamia nuts. If the total seed set on a tree is low, then nuts from inbreeding, or self-pollination, may mature. But if the tree sets a large crop, only outcrossed nuts are retained and the tree selectively aborts the inbred fruit.

These studies raise a question: Why do plants produce more flowers than they can ever rear? Is this not a waste? Sexual selection may cause overproduction in hermaphroditic plants. The first plants to set seed fulfill the female function, and the high cost of pregnancy and seed-rearing may soon put an end to fruit sets. Many plants such as peas and beans have a hormonal feedback system that sends out signals as the bean pod develops. These signals shut down flower production so that there will be enough resources left to feed the growing embryonic peas or beans. That is why gardeners must keep picking their beans and peas while they are young and tender—to ensure that the plant will continue to flower and set new pods over a long season. However, sexual selection may override this strategy. If there is a high value and low cost for the male function of the flower—that is, pollen production and dispersal—the plant may continue to flower even though the fruits that form will be aborted. A successful pollen grain will transmit exactly the same number of genes as a successful seed, so it may be worth the price of abortion to continue flowering and reaping the genetic benefits of pollen dispersal.

Abortion may also be selective according to the sex of the offspring. In a species with a highly polygynous mating system, usually only older, large, healthy and well-fed males can hold mating territories, yet virtually all females, regardless of their vigor, are mated. Under these conditions, a female that is not in prime condition and has limited maternal resources can maximize her reproductive success by selectively investing in daughters. The reverse tactic will work for females in peak condition. A shift from producing daughters to producing sons may occur over the lifetime of the female as she grows and accumulates more resources for reproduction. This sex-ratio adjustment strategy seems to explain the pattern of selective abortion in rodents. When female mice are given a nutrient-deficient diet, their litters are reduced in size and mostly female, because during the pregnancy, they selectively abort and reabsorb the male embryos. This may continue after birth. When food is limited, female wood rats with young selectively starve their sons and concentrate on feeding their daughters.

The tactic of overproduction followed by abortion is part of the strategy ecologists refer to as bet-hedging. An unpredictable environment, good or bad weather and the risk of attack by flower- and fruit-eating insects and disease mean a parent is always uncertain as to the number of offspring it is likely to be able to rear in a season. In good years, almost all the flowers may be used, while in bad years, most may be aborted. Selective abortion, then, provides a way for adjusting investment according to environmental vagaries.

The fact that embryos seem to compete for maternal nutrients means there are likely both costs and benefits to sibling rivalry that can lead to the death of offspring. Many raptors such as hawks and other birds such as herons and egrets have a Cain-and-Abel method of child-rearing. They lay two eggs, but normally, one chick kills its brother or sister by pecking it, starving it or shoving it out of the nest before it can fly. Since a sibling is just about as close a genetic relative as one can get, the pressure selecting for siblicide must be severe. It appears that the parents of these birds can normally rear only one offspring per season. Laying two eggs is also a kind of bet-hedging. In a very good season, they may actually rear both offspring. Much of the fighting between siblings is over food, and if food is superabundant, fighting might be reduced. Even if this were not the case, laying two eggs still covers the possibility that one of them may not hatch or may be damaged, defective or otherwise destroyed.

Theoretically, siblicide could be a case of parent-offspring conflict. It might be in the interests of an offspring to grow as large as possible even if it means killing a sibling. The children of a sibling are worth only half what an individual gains genetically from producing its own children. On the other hand, parents might maximize their genetic transmission by producing two medium-sized fledglings rather than a single huge one. In the case of brood reduction in birds, the process appears to be under the control of the parent. The parents produce eggs asynchronously, which automatically results in large size differences in the offspring. The youngest bird is destined to die except under exceptional circumstances. Snow buntings in the Arctic may lay a clutch of six eggs over several days, resulting in a variation in size. Yet when food is limited, they do not attempt to apportion it according to size, feeding the smallest more often. They feed according to who solicits most vigorously. These are the older, stronger and more conspicuous birds. So in bad years, several small chicks starve, evidence that the deaths are the result of parental bet-hedging—children are sacrificed as an adaptation to an unpredictable world.

In species in which embryos are fertilized and begin developing synchronously, there is scope for siblicide that would allow for brood reduction. In sand sharks and some salamanders, for example, the young hatch within their mother's body and begin to cannibalize one another while still in the womb. If growth and its attendant cannibalistic advantage are genetically determined, this sort of culling may be adaptive for the parents.

The thesis of this chapter is that abortion is a common female reproductive strategy used to control the trade-off between the quality and the quantity of the offspring. But this sort of manipulation is not limited to females. The significant point is that abortion involves destruction of one's own offspring. Normally, males have little opportunity or motive for killing their progeny. However, whenever male parental investment is high, the male is confronted with the same problem as the female: how to apportion resources. Male abortion occurs in fish such as the three-spined stickleback, *Gasterosteus aculeatus*. These fish build a nest and guard the eggs they receive from females. A male may receive the clutches of several females according to

Hellbender salamander: The benefits of dining in

how successful he is in courtship. A male may increase the total number of clutches he fertilizes over the season by eating some of the eggs he has fertilized as food. If he leaves to forage, his nest is soon raided and completely cleaned out, so he is under pressure to pay a short-term price for aborting (eating) his offspring in order to maximize his total long-term production. Hellbender salamanders, *Cryptobranchus alleganiensis*—those huge salamanders of Allegheny Mountain streams—apparently also eat some of the eggs they guard.

Abortion makes much adaptive sense for plants and animals faced with an unpredictable resource base and genetic pool. Yet for humans, it is a paradox. It makes less sense for low-fecundity organisms with plenty of scope for mate choice and rational resource choice. Perhaps that is why some societies are unforgiving of the strategy when it is employed for individual benefit. In the constantly warring and competing tribes that gave us religions such as Judaism, Christianity and Islam, there may indeed have been selection on groups to maintain high fertility levels in order to compete and to replenish the depletions caused by intertribal strife. Respect for individual rights as a means to promote the functioning, security and solidarity of a community may have also led to cultural selection for control of abortion. But despite this, abortion remains a large part of the human reproductive strategy, which testifies to the strength of individual self-interest. Abortion remains the world's most important method of birth control.

Abortion is a paradox that is difficult to accept because it says less is more. Many find it hard to believe that abortion and infanticide were used and are still used by some cultures and some women to raise their reproductive success, not reduce it. In affluent industrialized cultures with low infant mortality and social support for orphans and neglected children, it becomes easy to confound the difference between being and potential. In every pair of sperm and egg cells, there is the potential for a new human individual, yet when they combine, they must meet a long series of challenges before they can realize that potential and actually become a human being. Affluence and civilization have eliminated some of these obstacles, but in the past, the right to life was not given—it was something taken, fought for. Parents—not society, sex cells or embryos—made the decision, and that decision was made according to the simple logic of natural selection.

The logic behind the paradox of abortion explains why a healthy twin was discarded, why the malformed were destroyed. It explains why the very old and the very young were eaten during times of famine. It explains why, when under stress, aphids selectively reabsorb the tiniest embryos first or why a wasp colony in the grip of famine eats the eggs of the colony, then the larvae and finally the pupae. The pattern in this destruction is clear. Lives are valued according to the reproductive and genetic return they are expected to yield. Part of the human tragedy is that we have discovered the desire but not the means to transcend this harsh calculus on a small and crowded planet.

CHAPTER 8

Female Versus Female

"I know this is not exactly a burning question, but my feeling is that if female birds sing, then people shouldn't keep on saying they don't."

Ida Geary

CoEvolution Quarterly, Spring 1982

The greatest adventure story I have ever read is a true one. It is the story of Helen Valero, a member of a family of Brazilian campesinos settling in the wilderness region of the upper Río Negro. In 1937, when Helen was a young girl, the Valeros were attacked and Helen was kidnapped by the Fierce People, a band of Yanomamo Indians.

Stealing women is a Yanomamo tradition. Perhaps that is why Helen was kept alive. In a raid, Yanomamo generally kill whatever males they can and are not above bashing out the brains of even small male infants on the principle that a dead male can never avenge the raid or steal their women. Yanomamo males are often polygynous, maintaining a household with several wives. This means that several males within a village are without wives. Some anthropologists who study the Yanomamo credit both extreme competition for women and the tendency to prize ferocity and aggressiveness as the sources of intervillage strife and warfare. Males demonstrate their ferocity and achieve social status through a variety of gory chest-smashing and head-clubbing contests as well as through prowess at warfare and hunting. The machismo produced by male-male competition is certainly spectacular, but it is not the only form of reproductive competition in this society.

After Helen Valero had grown up and become the wife of a Yanomamo warrior, she was stolen again and marched by her captors to another village (*shapuno*), where she was destined to become the wife of one of the raiders. The resident wives were not glad to see Helen and her fellow female prisoners. Here is Helen's account of their reception. "Every wife of the *shapuno* said to her husband's female prisoner: 'Now you'll do as I say. You'll have to gather wood for me and water in the *igarape*. If you don't do it, I'll beat you.' The wife who was jealous knew what her husband wanted to do with the other woman, waited for them to go out, cut down the woman's hammock and threw it on the fire; then she looked for the *urucu* with which the other woman painted herself and threw it away. When she met the other woman coming back with the man from the *roca*, laden with bananas, she beat her with a stick.... The man looked on quietly. The wife hit the woman hard on the head; blood spurted out. Then the man took a big stick and gave it to the other woman."

Here is a clear case in which the females' reproductive strategy runs counter to that of the males. Polygyny is generally disadvantageous for women. It means that wives must share whatever resources the male has to offer. In the case of the Yanomamo, the wives must divide the meat from the hunt and the fruit of the husband's labor in the garden. Little wonder, then, that wives jealously oppose the addition of a new wife to the household.

Polygyny may cost the average female, but it benefits the male: The male's reproductive success will rise as he adds more females to his harem. For females, however, the reverse trend holds. This pattern has been established for a variety of animals, and it seems to hold for human societies in which polygyny has been studied as a reproductive strategy. An anthropological study of a highly polygynous society in Sierra Leone found that more than half the males were polygynous, and some males had as many as half a dozen wives. Monogamous males had an average of two children, those with two or three wives had an average of three to four children, and those with four or more wives had about seven children. Females experienced the reverse trend. Women from monogamous households had roughly two children, but as the number of wives in a household increased, the average number of children per wife dropped to less than one. Not only did the birth rate per woman drop as polygyny increased, but infant survival also decreased. This suggests that polygyny lowered the amount of parental care available per child.

In any human culture, polygyny and fertility will be confounded by a hugely complex set of variables. Nevertheless, a study of polygyny in

19th-century America that was carefully controlled for many variables found the same general trend. The pattern is also seen in other mammals with a less complicated social life. Polygyny is the most common breeding system of yellow-bellied marmots, those large ground squirrels common in the alpine meadows of the Rockies. A male's territory may contain from one to four females to which he has exclusive sexual access. A careful study has shown that the reproductive success of males and females—that is, the number of offspring produced—is inversely correlated. A monogamous male has the lowest reproductive success, while the reproductive success of a female with a monogamous mate is four times greater than that of a female in an extremely polygynous situation.

The reasons for a male to seek polygyny are obvious. But why should females ever be part of a harem? The most obvious explanation is that male territories vary greatly in quality. A female may get more out of a good polygynous territory than a poor monogamous one. Anthropologists have used this argument to explain the extreme polygyny of Australian Aborigines; the great variation among males with respect to age, status and power results in females choosing a polygynous but powerful husband over an unmarried but less influential male. And polygyny can be somewhat stabilized if the wives are sisters. Sisters are more likely to cooperate because of their high degree of genetic relationship. It may also be that if females enter into polygyny sequentially, then each has the potential to move up in the hierarchy of wives. As the oldest wife grows older, the second-oldest female may gain in influence and reward. But whatever is said about the possible benefits of polygyny to females, females ought to oppose it. Whenever males have something more than sperm to offer, we expect females to compete for male resources and, in fact, for any resources that are scarce to females.

In spite of this expectation, little has been written on competition among females. Dozens, more probably hundreds, of ecologists and their graduate students have spent tens of thousands of hours wading through cattail marshes measuring water depth, food availability, nest heights and many other variables in an attempt to demonstrate why female red-winged blackbirds prefer polygynous territories. This has been going on for several decades, yet only in 1984 were papers published showing that females actively opposed the settlement of other females in their territories. Female redwings even possess a distinct aggressive song, clearly audible, that is used against other females.

Another female vocalization which has been studied is that of the Mueller's gibbon, *Hylobates muelleri*, a monogamous ape of the rainforests of Southeast Asia. Gibbons are known for their swinging abilities—they spend almost all their days dangling by their arms and flinging themselves hand over hand in great leaps through the canopy. Their other outstanding characteristic is, according to J.C. Mitani in *Behavioral Ecology and Sociobiology*, their "loud and complex calls of considerable purity in a stereotyped manner, which capture the spirit, both joyful and melancholic, of the jungles of the Far East." The article goes on to say that "these beautiful calls, given mainly as duets, serve to maintain the pair bond and to exclude neighboring groups from the territory of the monogamous family group."

This all sounds harmonious and nice, and perhaps the duet does serve that function. But females also use song and respond to song according to their own interests. Female Mueller's gibbons sing in duets with their mates. Males sing like birds in the early morning; their songs are a tour-de-force proclamation beginning as a series of simple syllables and becoming progressively more complex until, after an hour, the song has evolved into a series of elaborate quivering trills audible more than half a mile away. Females do more than accompany this performance. They also sing solo. Interestingly, the two sexes respond differently to solo performers of different sexes. Male and female will both react to the territorial song of a strange intruding pair by cooperating to drive them out. But if a solitary female sound is heard, it is specifically the female that moves off through the forest to repel it. The investigator who studied this behavior concluded it was the aggressive response of the female to the song of solo females that forced monogamy on the male gibbon.

I cite this study because it is a rarity. Male-male competition is the subject of reams of print, yet only a small amount of work on female-female competition has been published. There seems to have been a tendency to overlook the occurrence and importance of female aggressiveness. In a paper in *Nature*, A.J.F. Holley and P.J. Greenwood reported on the courtship antics of the brown hare, *Lepus capensis*. In spring, these lagomorphs display behaviors responsible for the expression "mad as a March hare." They can often be seen chasing each other and delivering sharp cuffs, boxing one another's ears and faces. Previous studies more or less assumed that these boxing matches were

Brown hare: Females duke it out

male-male fights over access to females. But careful videotaped studies revealed that most of the boxing was done by females cuffing away unwanted male suitors. It may be that females have generally been underestimated as competitive and aggressive tacticians because the techniques they use are longer-term and less overt than pugilism.

The theoretical rationale for the focus on male-male competition and female choice stems from a famous experiment published in 1948 by geneticist William Bateman. Like most of his kind, Bateman worked with fruit flies competing within a universe consisting of a glass vial filled with artificial food. Bateman set up matings in which he gave males and females access to different numbers of mates. He used flies in which each individual and its progeny could be recognized by a distinctive genetic marker such as eye color. He found that, as expected, the reproductive success of a male was directly proportional to the number of females available. More important, he found that variation in male reproductive success was four times greater than variation among females. Evolution and natural selection act on variation. This led Bateman to conclude that, generally, there ought to be "an undiscriminatory eagerness in males and a discriminatory passivity in females," since evolution would favor the male that achieved the most copulations and tried the hardest. It would then follow that since all females could be inseminated, the most coy and discriminating would be inseminated by the most active and presumably fittest male.

But females can do more than passively discriminate. It may be that the focus of reproductive competition among males is often on gaining access to receptive females. Under these criteria, females may not compete. However, female reproductive success depends on more than just insemination; it depends on the resources required for egg production, pregnancy and the rearing of offspring. Selection operates within the sexes; a female's evolutionary success must be measured according to how well she competes with other females. Accordingly, we expect that when females live in close association, female-female competition will be important.

Social birds are particularly rich in female competition. Their commitment to self-advancement extends to killing the offspring of other females. The behavior was first described in subtle form for the groove-billed ani, *Crotophaga sulcirostris*, a thickset, black cuckoolike bird found in the American Tropics. Anis breed in groups composed of up to four monogamous pairs and some unpaired individuals. They forage beside large mammals such as grazing horses. Their tolerance for large mammals means that they can be easily observed by humans.

When Sandy Vehrencamp, an American behavioral ecologist, began studying the communal nesting groups of anis, she found that they were far from a peaceful assemblage of mothers. Females form dominance hierarchies. They are not especially violent but nonetheless have important effects on reproductive success. Social status determines which female wins the egg toss. Females laying eggs in the communal nest toss one another's eggs over the side. The dominant female does the most tossing and, in doing so, betters the others in reproductive competition. She must also achieve her success against the reproductive intentions of all males besides her mate; she is tossing their genes onto stony ground too. She nevertheless keeps doing it until she attains an advantage over female competitors.

Egg-tossing has evolved independently in a distantly related bird, the acorn woodpecker, *Melanerpes formicivorus*. Unlike most woodpeckers, this species is social, living in groups of up to a dozen individuals. These birds are common throughout the southwestern United States, wherever there are acorns and other nuts for them to store in their remarkable granaries. Each group has a storage tree into which the birds pound tens of thousands of holes where they wedge various nuts and seeds for use as winter provisions. The granary is carefully tended—the birds move nuts around to prevent mold and defend them from the depredations of foreign acorn woodpeckers. Food storage and cooperative defense of this centralized and concentrated resource provide a strong impetus for sociality. But an acorn woodpecker group is not a purely mutualistic entity.

Females toss eggs. They normally lay one egg a day when they reproduce, and they continue for roughly five days. The nest is communal, and one female has to begin the laying bout. Normally, the first egg laid is tossed. A nonlaying female usually removes it. She may carry it out to a limb and peck it open, and many members, including the female that laid it, may join in eating it, according to William Stacey, the ornithologist who has described much of this behavior. The destruction continues until the other females begin laying. Apparently, females cannot tell their own eggs from those of the others, so self-interest finally puts an end to the destruction. The winner is simply the female with the resources to lay the most eggs for the group to rear.

Males seem to play a minor role in the reproductive and social life of these birds. Other students of acorn woodpeckers have written that "competition and interference of communally nesting females are greater and affect the reproduction of groups significantly more than that of males."

Students of another social bird, the scrub jay of Florida, have observed that the females initiate most of the skirmishes related to carving out new territories. Females appear to be highly aggressive whenever they evolve the use of a limited and defensible resource. Cavities and burrows are good examples. In hole-dwelling birds such as toucans and parrots, females are often as large and brightly colored as males. The male-female pair must do battle with other male-female pairs to obtain a good nesting site, so it will pay a female to evolve the size and bright colors that signal status and fighting ability, even if it means a somewhat greater risk of predation. In eastern Ontario, Raleigh Robertson, a professor of biology at Queen's University, and his students kept track of what went on in nest boxes set out for swallows and bluebirds. They reported violent fights between females—to the extent of wounds that could cause death.

In primates, the correlation between female appearance and female competition may also hold up. I have yet to try to quantify a correlation between female coloration and competition, but I do know of some evidence that might make this worthwhile. Some of the most colorful primate females are tamarin monkeys such as the cottontop tamarin of the Panamanian and Colombian Chocó rainforest. Females are brightly patterned in blacks and whites, and they inhibit each other's reproduction. They leave scent marks according to well-developed dominance patterns within the group. There may be several females in the group, but only one at a time gets to reproduce. Since male tamarins provide some of the greatest parental care of all primates, it may be that females compete for male parental care and signal their status with scent and physical appearance.

Lest we forget, it is worth noting that such behavior forces males to be monogamous, whatever their inclination, and in many mammals, such as gray and ringed seals, burrowing rodents and canids, it is female-female aggression that results in monogamy and not high male parental investment.

Many other mammals show these patterns of aggression, especially when females live in burrows that are costly to dig. Female voles, burrowing squirrels and beavers are violently territorial and evict intruding females with great vigor. Their housing is so valuable, they often pass it along to their daughters and will wound any unrelated young females that try to establish themselves in the area.

The occurrence of this kind of female-female aggression seems to be correlated with sexual differences in the costs and benefits of dispersal. When males of social species are the dispersing sex, they tend to be shorter-lived than females. They have to fight their way into membership in a group. This is the pattern shown by many mammals ranging from ground squirrels to lions. Males fight more and die sooner than females. There is at least one social mammal in which the reverse pattern is found. African wild dogs live in groups composed of related males. It is the females that disperse. They are short-lived and compete fiercely. The object of their competition is male parental investment. Unlike most male mammals, wild dogs help raise the young by feeding them and defending them. The males are such good caretakers that they are able to rear a set of weaned but very young dogs without any female assistance. But a pack of males is able to rear only one litter at a time. When two females produce litters at the same time, the dominant female usually destroys the pups of the other.

The same pattern of behavior is also common in related canids such as wolves and coyotes. The murderous behavior is predicated on genetic self-interest. Lionesses that are closely related genetically—usually sisters and close cousins—do not molest the cubs of other females in the pride but care for them. Female wild dogs, because they disperse from different troops, are unrelated and therefore more competitive. This pattern is neatly shown by Cornell ecologist Paul Sherman's studies of the Belding's ground squirrel, *Spermophilus beldingi*, a native of the meadows of the Sierra Nevada Mountains of California. Females that have burrows in close association with related females such as sisters and daughters rarely fight. They cooperate in evicting unrelated intruders, and they warn one another of predators with alarm calls. If a female migrates to a new area where she lacks relatives, she commonly enters another female's burrow and destroys all the young. Sherman interprets this as the female's way of eliminating future competition from nonrelatives.

Burrowing females seem especially prone to this sort of violence. Female European rabbits, *Oryctolagus cuniculus*, will fight to the death for possession of a burrow. Males also fight, but their aggression is normally directed against males from other rabbit groups; females fight other females within the group. In many rodents, including eastern chipmunks and various mice and hamsters, the females will not tolerate the presence of either males or other females in their burrows and will violently evict intruders.

There are more subtle forms of female competition. In colonies of elephant seals, *Mirounga angustirostris*, the most-studied aspect of reproduction has been the impressive behavior of the giant harem-forming bulls. It has been established that females vary in social status and that their status affects the welfare of their pups. Offspring of high-status cows are bitten far less often than those of low-status mothers, and orphans are bitten three times as often as high-status pups. It is not known whether infant harassment translates into adult status, although studies of various primates suggest that possibility.

Many primate females practice psychological warfare on other females and their children, and in some primates, it is established that a mother's rank affects the social rank of her daughter. Yellow baboons, *Papio cynocephalus*, harass with some sophistication. Females form alliances and cooperate in harassing other females. These attack coalitions are able to suppress ovulation in their victims. They manage to suppress other females to the disadvantage of the male harem leaders: As more females are added to the troop, the male's reproductive return per female decreases.

Social competition among primate females is a drawn-out affair that extends throughout pregnancy and the lives of their offspring. It is harder to study because of that and because of its more subtle character. When observers first wrote about howler monkeys, they were inevitably moved to comment on the male. These monkeys don't really howl in the sense that wolves and coyotes do. They roar, grunt and hoot deep guttural noises that reverberate through tropical American forests. When Victorian naturalist Henry Bates traveled the Amazon, he recorded, in *Naturalist on the River Amazon*, the intimidating effect of these broadcasts: "Morning and evening, the howling monkeys make a fearful and harrowing noise under which it is difficult to keep up one's buoyancy of spirit. The feeling of inhospitable wilderness which the forest is calculated to inspire is increased tenfold under this fearful uproar." Males, which have the lowest, most basso profundo roars, are larger than females and have an enlarged vocal box. In short, for the species of howlers that are well known, such as the mantled howler, *Alouatta villosa*, males seem to play a familiar role. They fight, to the death if necessary, for control of a group. Males are conspicuous. But what of females?

More recently, howler monkeys have been studied by a large number of women: Katy Milton, Clara Jones, Ranka Sekulic, Margaret Clarke and Carolyn Crocker, to name a few. Crocker, in particular, has shown that female-female competition in red howlers is severe and involves physical combat. They harass each other and may even expel young females from the troop. A lone female may spend years dogging another troop before attaining membership, and she may have to fight her way in. It may be that females even kill one another's offspring. Summing up her work, Crocker stated: "This is a far cry from the picture of primate societies in decades past, when we imagined docile mothers did little more than raise cute infants. It's not the monkeys that have changed; it's our perspective."

What we believe about the nature of males and females is very much a product of the time scale and focus of our observations. Male-male combat tends to be intense at mating season and involves attempting to mate frequently and sequester off-guard mates. It is often easy to observe, since it is all packed into a short and furious rut. But where males may maximize their reproductive success by maximizing their copulation rate and the number of mates, females must maximize the amount of food they can acquire and the quality and number of offspring they can rear over their lifetime. This is a drawn-out task that goes on year after year in many mammals and other groups. Thus a different scale of measurement is needed to detect it.

Anthropologist Lionel Tiger wrote the best-selling book *Men in Groups*, a work that drew heavily on the primate literature. Tiger's argument was that male-male interactions, dominance hierarchies and a concept called "male bonding" had led male primates—human males in particular—to become political animals. Female competition was not considered significant. This led Tiger to conclude that "it may be unwise, therefore, to be optimistic or even sanguine about the possibilities that females will soon stimulate much change in the social subsystems and systems at the root of war as well as happier actions.... Women do not form bonds. Dependent as most women are on the earnings and genes of men, they break ranks very soon." Whether we attribute this attitude to Tiger's own particular view of the literature on primate behavior or to the actual state of the literature itself does not matter. Now his conclusion must be rejected. The strategies of female competition may be less familiar and more subtle than those of males, but they are there if we look for them.

CHAPTER 9

Milk and Honey

"The exact number of Nymphs is unknown.... Their name may simply mean 'marriageable woman.'... The ancients made them offerings of honey, olive oil and milk."

Jorge Luis Borges
The Book of Imaginary Beings

Grease is good. That is the message running between tongue and brain whenever a hungry person bites into barbecued ribs or French fries. It is the same sensation of pleasure that honey or anything sweet elicits from our taste buds. There is nothing inherently delicious in the molecules of honey or French fries. Our sensory system has simply been designed to perceive them as such. For those with a weight problem, this must seem a cruel and curious design: The things we are supposed to eat little of are the things that taste the best. We are programmed to overeat, a design that is now maladaptive for affluent humans but one which made good sense just a few millennia ago. For most of our history, refined oils and sugars did not exist. Concentrated calories in the form of honey and animal fat were scarce items that had to be hacked out of trees or run down and speared. People were thin, and fat was a sign of success.

Our literature is laden with images that link fat with human fertility. The biblical book of Genesis, a tale of human fecundity, uses "the fat of the land" as its image of prosperity. Homer's *Odyssey* speaks lovingly and frequently of sacrifices to good fortune consisting of livestock distinguished by a large quantity of fat. In the *Vinland Sagas*, which commemorated the Viking voyages to America—expeditions occasioned by population pressure in Iceland and Scandinavia—there is a scene in which Thorvald, the expedition leader, is skewered by an

Indian arrow: "Thorvald pulled out the arrow and said, 'This is a rich country we have found; there is plenty of fat around my entrails.' " The farther back we go into human art and literature, the more important fat is as a symbol. The earliest tributes are the Venus figurines, obese caricatures of women with huge breasts and buttocks, symbols carved some 30,000 years ago by the mammoth-hunting peoples of Europe and Asia. Art historians have debated the religious and artistic significance of these carvings, but their biological significance seems obvious. They indicate that our ancestors were well aware of the association between fat and human fertility, a relationship which may account for the strange physique and beauty of the human female.

The relationship between fat and female fertility has been quantitatively explored by Rose Frisch of the Harvard Center for Population Studies. She has demonstrated that females need to accumulate a certain amount of fat before they begin to menstruate and before they can become pregnant. This explains why, as cultures become sedentary and affluent, the average age of menarche, or first menstruation, drops. In the United States, women now become physically mature three years earlier than they did a century ago.

In affluent societies, menarche occurs at age 12 or 13 on average. It appears that this age is determined or largely influenced by economics. A Harvard-conducted comparison of the age of menarche in mothers and daughters showed that there was no difference between them when they were both members of the same socioeconomic class. However, when daughters rose above their mothers' economic class, they matured earlier. Presumably, they matured earlier because they had less physical labor to do and more food to eat and therefore got fatter earlier. It has been known since the Renaissance that urban women mature earlier than their rural counterparts. The converse is also true. When women are subject to starvation or for some reason become emaciated—from anorexia nervosa, for example, or intensive jogging—they cease menstruating.

Frisch has calculated that a woman must acquire roughly 17 percent body fat for menarche to occur and 22 percent body fat before she can become pregnant predictably. Getting fat enough to become pregnant is no longer a North American concern, but in nomadic cultures such as that of the !Kung of the Kalahari Desert and in subsistence groups such as the New Guinea highlanders, the average age of menarche is as late as 16 to 18. !Kung women average only 20.6 percent body fat and have a low fertility. Conception may not occur for four or five years after menarche.

Anthropologists have data to support the idea that in our ancestors' hunter-gatherer life, female fertility was limited by the ability to get fat. The nomadic !Kung groups have a seasonal body-weight fluctuation. After the season of good hunting, body weight peaks. The birth rate peaks nine months later. Since copulation rates do not vary seasonally, this suggests that food intake determines fertility. Sedentary !Kung, who maintain milk-producing livestock, show a reduced seasonality in birth rate, which is in keeping with their more aseasonal food supply. In subsistence cultures, a woman may wait half her reproductive life to reach menarche, conception may be another four or five years later, and menopause comes early. This probably explains the repeated appearance of the custom known as "fattening the bride."

In coastal Sierra Leone, girls at puberty are stuffed with calorie-rich food and vegetable oils. In Ghana, the puberty ritual for females includes feeding them oily stews of meat, which the people claim make the women fat and give them the oily skin they find beautiful. Anthropologist Carol MacCormack notes that the Tamils of Sri Lanka seclude young women and feed them on "raw egg, eggshell, gulps of margosa oil and a dish of steamed rice, black gram, grated coconut and sesame oil. They are also fed vegetable curries made with sesame oil, cakes, sweets and puddings."

The fattening of females is also hormonally programmed. Before puberty, boys and girls differ only slightly in fat, muscle and bone proportions. At puberty, boys start to gain muscle and bone, while girls gain fat, so women end up with roughly 30 percent of their weight as fat, twice as much as men. Exactly how these fat stores are used by females is not understood biochemically, but the energetic value is obvious. Frisch calculates that a woman with 28 percent body fat has just enough calories stored to sustain a pregnancy and three months of breast-feeding. A human pregnancy costs the woman 50,000 calories above her normal nine-month requirement. Lactation burns another 1,000 calories daily. As Darwin and countless farmers have noted, it is very difficult to fatten a lactating cow. This may be why so many cultures have historically regarded the birth of twins as a bad accident and cause for infanticide. The Enga of New Guinea, for example, believe that no woman can nurse two children without endangering the health of all of them.

The design of the female hormonal system still reflects the energetic costs of childbearing and rearing. It contains an elegant feedback mechanism that guards against the woman's becoming pregnant while still nursing a child. When a baby is born, the neural and hormonal sensitivity of the mother's nipples increases sharply. Sucking stimulates the nipple to release oxytocin, a hormone that causes the discharge of milk and triggers another hormone, prolactin, which regulates long-term milk production. The elevation of these hormones then affects the hormones that control ovulation. The more a woman breast-feeds, the less likely she is to ovulate.

R.V. Short, a British authority on reproduction, writes: "Breast-feeding is nature's contraceptive." This is not an effective method for well-nourished women who breast-feed only lightly. However, light breast-feeding is a modern phenomenon. Traditional !Kung women suckle their children dozens of times daily, and they sleep with them and allow them to feed at night. This continues for an average of 3.5 years and is no doubt responsible for the fact that without contraception, the birth spacing is 4.1 years for these women. An experimental study showed that breast-feeding five times or more a day for a total of more than one hour was enough to control ovulation completely. The point, then, is that the demands of lactation and pregnancy have been so stringent that natural selection has resulted in the evolution of an effective birth-control mechanism allowing women to space their pregnancies.

So far, none of this would suggest anything more than the fact that the female form, with its prominent breasts and buttocks, is a device evolved to store the fat needed for pregnancy and lactation. Some comparisons, however, show that the case is not so simple. Females of other mammalian species also undergo long and energy-demanding periods of pregnancy and lactation, yet they have not evolved a similar physique. Only our species has such enlarged breasts and buttocks. This has led biologists to speculate that sexual selection for mate attraction has led to the evolution of the human female figure.

The first explicit suggestion of this idea (to my knowledge) comes from psychologist Havelock Ellis in a 1905 book on sexual selection in humans. According to Ellis, the most basic and primitive standard of feminine beauty was one based on fertility, breasts and buttocks. "The most beautiful woman is one endowed, as Chaucer expresses it, 'With buttokes brode and brestes round and hye'; that is to say, she is the woman best fitted to bear children and suckle them."

The literature on hunter-gatherer societies is full of evidence for male appreciation of female corpulence. When anthropologist Allan R. Holmberg studied the nomadic Siriono of the Bolivian rainforest, his male informants told him that "a desirable sex partner—especially a woman—should be fat. She should have big hips, good-sized but firm breasts and a deposit of fat on her sexual organs." Brenda Grey, an anthropologist from Johns Hopkins University, reports that among the Enga of New Guinea, "a sleek, fat body is regarded... as a most important physical asset in a young woman. A thin girl is considered unlikely to make a good marriage.... Each girl undertakes a simple personal ritual which she carries out each day at dawn so that her body may grow fat and her skin sleek and smooth.... A fat woman is believed to live longer, and her children are more likely to survive."

Other subsistence cultures show a similar reverence for fat females. In some, the perception is so deep that the symbolic links are embedded in the language. When another anthropologist, Jules Henry, studied the Kaingang, a group of Brazilian nomads perpetually plagued by hunger, he noted the relationship between food and fertility: "The Kaingang are ridden by hunger and swayed by the strength of their libidinal ties, and they have in some degree overcome the one by building on the other. Food and sex have become so closely interwoven in the Kaingang mind that they even use the same term for eating as they do for coitus." The Tukano, a rather different tribe found in the region of Colombia's Vaupés River, are reported by Reichel Dolamtoff to have in their language the verb *vaimera gamatari*, "to hunt," the literal meaning of which is "to make love to the animals."

This is not merely a notion of nomads, males and anthropologists. Even Germaine Greer, once a feminist, writes in *Sex and Destiny* that "where most people live at subsistence level or below, fatness is a sign of success and wealth, but the different aesthetic has deeper roots still. The most prevalent notions of beauty are those of fecundity."

Critics of this scenario will raise as an objection the trend toward emaciated fashion models in Western culture or will follow Darwin, who thought of sexual selection and beauty primarily in terms of ornamentation such as tattooing, lipstick, lip plugs, presses for flattening the skull, high heels, bouffant hairstyles, teeth filed like fangs, incised scars, stretched necks and all the other ways humans have of altering their appearance. As Darwin pointed out, these kinds of beauty are culturally plastic, but that is not to say the traits which promote sexual attraction are entirely arbitrary. (I think Victorian modesty prevented Darwin from writing explicitly about the virtually universal attention given to female breasts and buttocks.) Fashion is more a signal of social status, wealth and acumen than it is a device to promote sexual attraction. In human society, status, kinship, wealth and attitude are all important considerations in mate selection. In this kind of system, any sort of fashion, even anorexia, might be culturally selected if it conveys identification with an upper social echelon. Indeed, in our society, in which the masses are grossly overweight, it is almost an a priori expectation that fashions pitched at a wealthy minority ought to emphasize the svelte. (Incidentally, by the standards of hunter-gatherers, an American of average body weight would be considered outrageously and possibly grotesquely fat.)

But our physique and our ability to assess the quality of a body began evolving long before clothes were ever thought of. Fashion and the standards of cosmetic beauty are capricious precisely because they have had so little impact on human reproductive success. I was interested to read an article on beauty by Steven Shiff in *Vanity Fair*. Although he was apparently unaware of the biological arguments about beauty, he concluded: "Even though styles in female dress and female models have changed over the past three decades... fashions in female nudity have remained frozen."

One of Konrad Lorenz's most creative studies was an analysis of the morphological features that convey juvenility, features such as a big, rounded head, indistinct receding chin and large eyes which tell us a dog is a puppy and not an adult. Those same traits elicit the word "cute" from people looking at infants, be they of humans or harp seals. Steven J. Gould has creatively applied the analysis of the infant schema to the development of the cartoon character of Mickey Mouse. He points out that Mickey's behavior has coevolved with his facial features, starting off initially as a long-nosed, beady-eyed wise guy before Disney cartoonists gradually softened him into a big-headed, wide-eyed nice guy with a childlike personality. Like Lorenz and Darwin, Gould argues that "the abstract features of human childhood elicit powerful emotional responses in us, even when they occur in other animals. I submit that Mickey Mouse's evolutionary road down the course of his own growth in reverse reflects the unconscious biological discovery by Disney and his artists." By the same logic, the enduring male interest in breasts and buttocks might suggest that the female form elicits an innate male response; its special geometry has been used by Ice Age carvers, by the inventors of bodices, bustles, corsets and brassieres and by Hollywood producers and pornographers alike.

Some dispute this. Anthropologist Donald Symons claims there is no evidence that any features of human anatomy were produced by intersexual selection. Human physical sex differences are most parsimoniously explained as the outcome of intrasexual competition (the result of male-male competition). Symons accords females little role in mate selection other than using a healthy-looking skin or copulation as a bargaining chip. He believes it is competition and combat between males that produce the physical dimorphism of males and females. But this says nothing about why women look so different from other primates and other mammals.

To understand the appearance of the human female, we must think a little more about the evolution of men. Humans are among the least sexually dimorphic of all the primates. Males have even lost the large, fanglike canines used by apes and monkeys in male-male combat. This hardly suggests intense selection and a mating system based on success at male-male combat. It is equally plausible that the more muscular and larger male physique in humans reflects selection for hunting ability rather than combat.

In hunter-gatherer societies, hunting success is a large determinant of a male's mating success. Holmberg reports the story of a Siriono man who was a terrible hunter. He lost his wives and was ridiculed because of his failures. Holmberg taught this man how to use a shotgun. His meat production increased dramatically, and eventually, he ended up with two wives. (It is not known what befell him when Holmberg took his shotgun and returned to the United States.)

Hunting is not a skill that depends particularly on large size, but a certain amount of upper-body strength is required. Perhaps that is why men are only slightly bigger than women. In virtually all cultures, men hunt and women do not. Selection for intelligence and large brain size in humans has greatly enlarged the female pelvis, so women are less efficient runners, especially if they are nursing a child. In ancestral terms, then, this division of labor makes sense. It has nothing to do with males being killers and fighters. This fact and some other data suggest that human sexual dimorphism may be more related to male hunting than to male fighting. A study of five North American native groups that varied in how far they had shifted from hunting to agriculture as a way of life showed that the more a tribe depended on hunting, the greater was the sexual dimorphism in skeletal structures.

Hunting is a secondary issue. The essential point is that men probably achieved their reproductive success not simply by fighting among themselves but by increasing their ability to gather resources such as meat. Men provide relatively huge amounts of parental investment in both their wives and their children, and for most of history, this may have been the main determinant of their reproductive success.

Whenever males invest heavily in their mates and offspring, they become a resource for females. We expect females to compete for males who possess more resources than usual. Thus many anthropologists and biologists take the reverse view of Symons: They believe that the human female competed by advertising her fecundity with a conspicuous display of fat reserves in her breasts and buttocks. In many mammals and birds, males invest less and are the gaudy, attractive sex, but in humans, females are the physically ornamented sex. A man could support only a limited number of women, and he would be under selective pressure to choose wives who would yield the highest reproductive return.

Humans have for a long time taken our peculiar system for granted. English literature is full of poetry such as Keats' "Ode to a Nightingale," which assumes that the sweet voices come from a female. Milton writes in *Samson Agonistes*: "But who is this, what thing of sea or land? Female of sex it seems, / That so bedeck'd, ornate, and gay, / Comes this way sailing." Only a human male could think this way—and only one with a limited knowledge of natural history, for in nature, it tends to be males that are bedecked and gay.

There is an irony here: The unknown selective pressures that caused men to assume an increased share of the burden of parental investment may also have generated the physical stereotype of woman as sex object. The hypothesis that the male taste for fat, fertile women has affected the female shape is not easily tested. Those who find it offensive may dismiss it as mere speculation, and they may take some solace in the fact that in our society, where the calorie is pariah and fat unfashionable, and in a world groaning with the weight of human fertility, such a taste is now well behind the times.

CHAPTER 10

What Good Is a Bastard?

"Cecrops, the first king to recognize paternity, was also responsible for instituting monogamy."

Robert Graves
The Greek Myths

Human females are sexually unpredictable. That is a relative statement made in reference to other species. Women have abandoned the use of estrus—that time when sexual receptivity, advertising and the likelihood of ovulation and impregnation come to a peak. Most mammalian females come into heat, or estrus, at regular and predictable times and in an ostentatious manner. Any owner of a cat or dog knows just how predictable other mammals can be. The female cat scrapes and rubs herself conspicuously around the landscape, and a canine bitch in heat gives off a scent that can lure male dogs from miles away. There is no ambiguity about the state of these females: Either they are in heat and mobbed by males, or they aren't and are utterly ignored.

The human female, by comparison, is inscrutable. A male cannot sense when a female is about to ovulate. The woman may signal receptivity, but even she is usually unaware of the moment of ovulation. She gives off few cues about her ovulatory cycle. Given the important consequences of this for both sexes, estrus and our apparent lack of it are biological curiosities. Why do human females find it advantageous to be inscrutable?

Estrus is advertising. It is also manipulation. We may go a long way toward resolving some of the mystery surrounding human estrus loss by thinking of estrus as just part of a range of signals females may send to males, signals designed to manipulate their behavior.

To understand the human position on this spectrum, we must compare ourselves with our relatives. Primates span the full spectrum of estrus-advertising behaviors, and they also employ a wide variety of mating systems, ranging from rigid monogamy to rampant promiscuity and extreme polygyny. The first step is to ask how estrus-signaling is correlated with a species' mating system. Intense female signaling seems correlated with a troop organization based on promiscuous matings among several males and females, as in common chimpanzees. Female chimps have regular menstrual cycles some 35 days long. Half the time, they are in estrus, a condition they advertise by lurid pink and red swellings of the skin around the groin. Normally, they copulate with several different males repeatedly during estrus.

Sexual swellings are known in other primates, and they seem correlated with a promiscuous multimale mating system. Zoologists Paul Harvey and Tim Clutton-Brock, who first documented this correlation, suggested that it was a form of female incitement designed to enhance male-male competition for females. The more males compete, the more they can winnow out less fit individuals. Primatologist Sarah Hrdy points out several problems with this interpretation. The form of competition in this case is sperm competition. The male that mates the most will sire the most offspring. But that may not be a good indicator of overall male quality. Females already have access to considerable information on male social status and age because they live with them all the time. So this form of incitement may be unnecessary. Hrdy suggests that it may simply be female efficiency at work. By advertising, she places the onus on the male to accomplish his task before or at the expense of his rivals. But this still raises the question: Why should a female submit to the bother of so many extra copulations?

Another possibility recognized by Hrdy and other primatologists is that sperm competition decreases the certainty of paternity for males. This makes it difficult for a male to discriminate against the offspring because it may be his own. Male discrimination can extend to killing unrelated infants in order to bring the mother back into estrus, so it would be advantageous for a female to conceal paternity whenever male parental investment is dispersed among a group of females. Females might thus force males to not only forgo infanticide but to actually increase the total amount of parental investment. In most cases in which paternity certainty is low, males tend to invest little. But in a social troop with a relatively stable composition, paternity certainty will never be zero. And where parental care is necessary, it would not pay a male to withhold parental care entirely—the offspring would then be certain to suffer. Rather, he should spread it among all the offspring of females with whom he has copulated.

David Taub, a primatologist at the University of California, explains female promiscuity in Barbary macaques as an adaptation that causes increased male parental investment. These natives of North Africa are extremely active during female estrus. A female may copulate for several days at the rate of almost four times an hour, and she usually mates with every adult male in the group. This does not mean that males forgo parental investment completely; it simply means that it is less focused. Males carry and defend the infants of more than one female. Their investment is spread around probabilistically. The same argument has been used to explain why female acorn woodpeckers that associate with groups of dominant and auxiliary males are also promiscuous. It was found that males which copulated with females, even though they were not closely pair-bonded, helped care for nestlings. Resident males that had not copulated offered no help.

We may also think of increased female copulation rate as selection for honest advertisement on the part of the male. Increased rates of copulation give monogamous females a true test of male physiology. If copulation and sperm production are physiologically costly, they are nonbluffable cues. British zoologist Brian Bertram, who has studied the reproductive biology of lions, makes the following estimate of how much a lion must pay in copulations for genetic representation in the future: "Assuming that lions mate every 15 minutes for three days, that only one in five three-day mating periods results in cubs, that the mean litter size is 2.5 cubs and that the mortality rate among cubs is 80 percent, then a male must mate on average some 3,000 times for each of his offspring reared to the next generation." Certain cats and some rodents require the stimulation of multiple copulations in order to effect the neural and hormonal changes required to initiate pregnancy. One ejaculate may be enough to fertilize a female, but it may require repeated copulations to stimulate egg release and implantation. This may be a means of assessing the territorial and social status of a male. A multiple copulator is less likely to be a transient intruder that will offer no care and expose the female to the risk of male infanticide if he leaves, and multiple copulations may testify to his physiological competence.

Female choice of honest male advertisement could be adaptive under conditions of promiscuous polygyny and monogamy when copulation is costly for males. The difficulty of producing multiple ejaculates would exclude unhealthy males from the mating tournaments. In addition, copulation in a social troop usually is a conspicuous affair affected by dominance relationships. Subdominant males can rarely copulate unnoticed. Thus by copulating frequently, a female increases the likelihood that she will be mating with a dominant individual. In effect, she forces honest advertisement on the would-be fathers of her children.

In monogamous relationships, this female strategy would be of no use. A male is assured of his paternity and the return on his investment by mate-guarding a single female. Estrus would not help a female maintain male proximity and its attendant benefits of guarding and child care, predator detection, and the like. Night monkeys, titis and marmosets share with humans a tendency toward monogamy and a high level of male parental care. Marmosets have twins, and the father often assumes complete responsibility for holding and transporting the newborn infants, which he passes back to the mother only for breast-feeding. This enables the mother to give birth to larger offspring than would be possible if the male did not provide such extensive assistance. Marmosets also have reduced estrus-signaling.

Such signaling may increase the efficiency of insemination, but I doubt that a time factor of this sort is important in the lives of most females. It would be only for male convenience. In fact, pronounced estrous signals would make it easier for the male to philander. If he could assess when a female could not get pregnant, he could leave her in pursuit of estrous females and thus offer less parental care.

Similarly, in stable harem mating systems such as those employed by gorillas, estrus ought to be reduced. One dominant male has exclusive access to the receptive females, so advertisement seems irrelevant for females. Again, a lack of it might force the harem master to tend his females more carefully. There would also be straightforward physiological costs to advertising that could be avoided, but male harassment may be the most important cost.

Indiscriminate advertising may attract unwanted business. For primate females that have few chances at reproduction, pregnancy is serious, and nothing about it ought to be slipshod or opportunistic. A pregnancy at the start of the dry season, prior to dispersal or during illness, for example, could prove disastrous. Restricting advertisement is one way of controlling the timing of pregnancy. It leaves the female the option of taking the initiative rather than making it profitable for males to initiate sex.

Concealment of estrus ought to be of concern to females whenever there is the opportunity for harassment by male rapists. Female orangutans are widely dispersed throughout the territories of adult males. Females forage alone and at low densities in the canopy of Southeast Asian rainforests. They get no assistance from the father of their children and little protection. The adult male defends his territory against intruders but cannot always prevent young, nonterritorial roving males from raping females within it. Birute Galdikas, a longtime student of orangutans, has observed many rapes, and all have been inflicted by young males. It may be that rape is reduced if a female shows no sign of being in heat. Rape involves a physical struggle with the female and some risk of detection by the territorial male. If a female is not sexually predictable, rape is less likely to occur, and if it does occur, it is less likely to result in pregnancy. Orang females show little sign of estrus, and it may be that their diffuse territories and lack of mate-guarding has made estrus loss adaptive.

Beverly Strassman, an evolutionary biologist, suggests that the risks of rape are manifest in the sensory physiology of human females. She cites physiological studies showing that as women approach ovulation, they experience significant improvements in their hearing and eyesight and even in their ability to smell musky male pheromones. According to Strassman, "This may have improved the ability of females to avoid rapists operating by stealth or ambush."

Donald Symons believes that estrus loss has more direct material benefits to females. He suggests that females trade sex to males in return for meat. His idea derives from the observations that male chimpanzees often share with females the meat they obtain by hunting and that they attempt to entice females into sexual consortships by means of food-sharing. Other authors have noted similar correlations between the resources human males offer females and the females' willingness to offer copulation in return.

Randy Thornhill has argued that estrus loss would enable a female to sneak copulations with a male other than her mate. This could increase the genetic diversity of her progeny—production of genetically diverse progeny is thought to be one of the main objects of sex in the first place. This would be especially important when males force monogamy on females by defending an important resource such as a nest site and excluding other males from their territory. If a female were sexually unpredictable, the territorial male's ability to guard her would be reduced. This may explain why multiple paternity is reported for some organisms such as bluebirds, which had previously been thought monogamous. A significant proportion of the males are cuckolded. In ground squirrels, multiple paternity may be much higher, most litters being sired by more than one male. Since male ground squirrels do not provide substantial parental investment, they may invest more in seeking copulations than in mate-guarding.

But some paternity uncertainty exists even in human societies in which the male makes a large parental investment in the children he believes are his. Studies of peoples as different as the Yanomamo of the Venezuelan rainforest and residents of the rural midwestern United States reveal that on the order of 10 percent of children were not fathered by the male who believed, and acted as if, he was the father.

Improved blood and antibody tests can now settle disputes of paternity, and responsibility disputes among humans often end in a surprising manner. One investigator did blood tests of 67 males who had been sued in court over paternity and who had conceded that they were indeed the father without going through blood testing. The follow-up study showed that 18 percent conceded paternity and accepted legal obligations for children fathered by someone else. This highlights the difficulty confronting males that must mate with inscrutable females. The female almost always knows her offspring; the male can only guess.

Several explanations for human estrus loss suggest that it is an artifact, an epiphenomenon generated by selection on other characteristics. One idea is that selection for increased female endurance required by bipedalism and nomadic and long-distance foraging required a rise in the muscle-building androgen hormones, which would reduce the expression of estrus. I find this weak, because other primate females that live lives of exertion do have estrus. There seems to be no compelling physiological reason for humans not to advertise estrus if it is to their advantage.

Along the same lines, it has been suggested that our upright posture made sexual swellings less conspicuous and noticeable. By this logic, women lost estrus advertisement simply because it was visually ineffective. But males would be more than willing to bend over if there were any useful message to be read.

Nancy Burley, in an article in *American Naturalist*, suggests that women evolved so as to conceal ovulation from themselves. She argues that a woman's fear of pregnancy and childbirth might reduce her reproductive rate. A woman's inability to predict her ovulation would raise her reproductive rate. However, I doubt such inhibitions could ever evolve. If there is any feature of reproductive behavior that characterizes human females, it is the effort to regulate pregnancy through contraception, abstinence, abortion and infanticide to achieve high-quality, well-spaced offspring, not ill-timed, unhealthy children. In all probability, too many pregnancies rather than too few have been the human problem. Human females, like the males, have been selected by social evolution to survive and transmit their skills and resources to their offspring.

Any explanation of human estrus loss ought to address our special tendencies; it is not enough to take refuge in historical argument. The need to produce intelligent, self-sufficient children has resulted in heavy male investment in child care. This investment need not be thought of as infant care per se but as the acquisition of material resources, social influence and knowledge that mothers and fathers can transmit to their offspring. Parents, both female and male, acquire these resources at the expense of reduced casual matings, freedom and mobility. But this parental investment is the male weakness that females exploit. In Shakespeare's *Love's Labour's Lost* is a song: "The cuckoo then on every tree / Mocks married men; for thus sings he, 'Cuckoo; / Cuckoo, cuckoo'—O word of fear, / Unpleasing to a married ear!" Through cuckoldry, a married, high-investing male may have his love's labor lost.

Estrus concealment must increase the potential threat of cuckoldry and theft of male parental investment. The male cannot assess when a female must be guarded. This strengthens a female's ability to force monogamy on a male. Sociobiologist Richard Alexander and his collaborator Katherine Noonan point out that if a man is uncertain of the reproductive state of his sexual partner, he will be forced to guard her. This means he will be less able to seek other mates. To the extent that he has parental investment to offer, he will be forced to invest it in the offspring whose parentage he is guarding. They further point out that monogamy will lead to less strife among males within a group. Many anthropologists and biologists believe that intergroup cooperation has been an important factor in human evolution. If certain groups or lineages are competing with each other, either directly through intertribal war—the common state of affairs in preagricultural societies—or through growth rate, then behavior that improves cooperation within the group can be selected for. Monogamy ought to lead to less male-male strife within the group than polygyny, in which some males have multiple wives and other males have none.

Rarely in nature is group selection as important a factor as individual selection, and we could discount this portion of the scenario but for the fact that estrus loss may actually be to the benefit of the average individual female and male. Only highly polygynous males would suffer from the transition to estrus loss and monogamy. The average or less fit males and most females ought to benefit by this trend.

Where social competition demands that males invest heavily in their offspring, males have only two means for greatly increasing their reproductive success: polygyny and cuckoldry. Polygyny is usually disadvantageous for a female; she can combat its use by increasing the amount of time her mate must spend near her and her offspring. Estrus loss, then, is a female's way of using the threat of cuckoldry as a weapon against polygyny.

This is a speculative argument as far as our own species is concerned. We have only anecdotes to go on. The argument will stand or fall according to how it fits the comparative data on other primates of varying estrus, parental investment and paternity certainty. Unfortunately, little good information on this for any primate species exists. It is doubtful that such a data set will ever be gathered before the natural habitats of the Americas, Africa and Asia are too greatly altered ecologically. But even if it is doomed to remain speculation, I may admit some small contentment. The tone of the literature on women and their biological oppression by men and nature is often justifiably angry. It is some relief to see a suggestion that female physiology has triumphed at the expense of the male.

CHAPTER 11

Orgasm and Inertia

"Wherever I found a living creature, there I found the will to power."

Friedrich Nietzsche

Thus Spake Zarathustra

Most animals seem appallingly lazy. For much of the day, they lie about dozing, sunning and generally goofing off when they could be busy looking for food or otherwise trying to get ahead in life. People whose impressions about wildlife come from television specials are in for a shock of boredom if they ever attempt nature photography or animal watching. For every shot of a group of lions pulling down a wildebeest, the camera crew spends hours, sometimes days, watching the lions loll about. That's one of the reasons the world is not overrun with people studying animal behavior, glamorous as it may seem when presented on a television screen. Many a graduate student has bailed out when he or she finds that a Ph.D. in animal behavior may mean spending hundreds, even thousands of hours watching animals sleep and scratch themselves and only a small amount of time seeing them do interesting things.

There is an obvious adaptive reason for this indolence. There is no point in hunting when your stomach is full or when there is no food to be had. Better to lie still and avoid having accidents and wasting energy. But total inactivity and contentment is not the answer. The laziness circuits of the animal must eventually be overridden by more inspirational circuits such as hunger and then further motivated by some rewarding pleasure that marks the successful conclusion of the inspiration. Meat, for example, undoubtedly tastes good to a cat, and the pleasure the cat gets from tasting it is part of an adaptive plan. It is the proximate mechanism, a built-in motivational device that serves the ultimate goal of acquiring the nutrients needed for reproduction. An animal, then, is an endless cycle of driving discontent, motivation, satisfaction, enjoyment, contentment and discontent again.

Orgasms are motivational devices. For males, the adaptive payoff is obvious. Males that are self-rewarded by orgasm in frequent copulation will in general have a higher fitness than those that lie around and sleep. Biologists, however, have had difficulty in applying this simple logic to females. The conundrum is this: If males are more than capable of inseminating all females, then a female presumably does not require a device that causes her to seek out many copulations. Indeed, since too many or inappropriately timed pregnancies could reduce the total fitness of a female, then orgasms might even be maladaptive. Moreover, for a long time, it was thought that only human females experience orgasm, and it was difficult to understand what function female orgasm would serve.

One biological anthropologist, Donald Symons, who has researched the female orgasm, concluded that the female orgasm is not an adaptation of females per se. It is merely a by-product of intense selection for the male orgasm. This selection and the neural information required by the act of copulation have led to a condition of genital sensitivity in both sexes. Given enough stimulation, Symons argues, any female primate can experience orgasm simply as a building and release of neural stimuli. In his words, "The human female's capacity for orgasm is no more an adaptation than the ability to read."

Symons concedes that "abundant evidence exists that—like other female mammals—women usually enjoy sexual intercourse." However, "female orgasm is an adaptation only if in ancestral populations, orgasmic females enjoyed greater average reproductive success than nonorgasmic females." One of the reasons Symons rejects the proposition that female orgasm is adaptive is that many of the previous arguments in favor of the idea depend on the dated notion of the pair bond. Symons quotes and rightly criticizes the arguments of several well-known biologists and psychologists because they are based on the assumption that the monogamous human pair bond selects in favor of the pair bond.

Desmond Morris writes: "As with both appetitive and consummatory behavior, everything possible has been done to increase the sexuality of the naked ape and to ensure the evolution of a pattern as basic as the pair bond." Eibl-Eibesfelt argues that the female orgasm "increases her readiness to submit and, in addition, strengthens her emotional bond to her partner." David Barash, a well-known sociobiologist, suggests that sex is a device for bonding male and female together because it raises the fitness of both partners and, further, that it may be "selected to be pleasurable for its own sake in addition to its procreative function. This would explain why the female orgasm seems to be unique to humans." It is easy to agree with Symons that this approach needs rejection. And yet we need not agree that the female orgasm is merely a by-product.

There is no evidence in favor of the idea that female orgasm and increased sexual activity are important in maintaining a pair bond. Mammalogist Devra Kleiman analyzed the phenomenon of mammalian monogamy and concluded that sexual activity "occurs infrequently and thus must play a minor role in pair-bond maintenance" and also that "there are no more intense sociosexual interactions in species exhibiting long-term pair bonds than in polygamous forms." This, of course, is a broad generalization drawn from a survey of all mammals. But it seems to hold up when applied to our primate relatives. In the monogamous gibbons, simiangs, marmosets and tamarins, sex between monogamous partners is not very frequent, while in polygynous multimale group species, such as common and pygmy chimps, copulation is very frequent. Dianne Harley, an anthropologist reviewing this idea, concluded that "continual sexual activity and attraction between a pair are not necessary preconditions for the maintenance of long-term bonds between males and females in the apes. It can be argued that the opposite correlation exists." Given this, it would be wise to abandon pair-bonding as the key to understanding female sexuality in humans and other primates.

The discussion of the significance of female orgasm is fraught with methodological and anthropomorphic complications. A Heisenberg principle is operating here: To demonstrate orgasm physiologically and experimentally involves wiring the subjects with electrodes in an exceedingly unnatural environment, and it may be difficult to measure orgasm under those conditions. And in the wild, our anthropomorphic limitations may make it difficult to assess whether orgasm is occurring. Can we equate the loud yells, apparently involuntary, of female baboons during copulation or the facial grimace and the clutching and reaching reaction of a female stump-tailed macaque toward the male as signs of orgasm? Is the pelvic thrusting of female monkeys, the display of their genitals and their masturbation testimony to their orgasmic potential?

Many primatologists believe female primates commonly experience orgasm, and many believe it must have a function. Katy Milton's ideas are typical of this view. She notes that "female behaviors may include (1) putting a hand back to the male partner, (2) glancing or grimacing at the male partner or (3) calling out around the time of male climax and ejaculation.... In some cases, the female exhibits an apparent orgasmic response around the time of male climax (e.g., rhesus monkeys, stump-tailed macaques, gorillas, humans). It seems unlikely that the clear occurrence of these female behaviors around the time of male climax and ejaculation is purely fortuitous. Rather, such behaviors are likely to serve some reproductive function.... Given the widespread occurrence of these behaviors, it does not seem correct to view female primates as passive partners in coitus; rather, female primates of many species appear to play an active role in soliciting and controlling male mating activity."

In the same sense that male orgasm motivates males to copulate, we might assume (perhaps incorrectly) that orgasm also evolved to increase a female's interest in, and thus her rate of, copulation. How could increased female copulation enhance fitness? The effect of increased female copulation on polygyny could take several forms. As noted for estrus loss, it may increase paternity uncertainty in multimale troops. Males would be less able to ascertain which offspring were their own and thus less able to discriminate selectively against unrelated infants. The net result would be to raise the average female's fitness. It has been suggested that this explains why lions copulate so frequently. An alternative suggestion about lions' copulatory excesses is that the willingness of females to copulate with several males for hundreds of bouts means there is little selective pressure favoring male-male combat for access to estrous females. Thus there will be less disruption of the pride, which disturbs females and increases infant mortality. Since prides are composed of closely related kin groups, this tactic might be favored over an alternative tactic of unruliness by males in a competing kin group.

Male lions will care for and defend infants. A female may manage to increase male parental investment if all the females play the same game. The stability of the strategy and the payoff to the individual lioness are enhanced by the fact that no single male monopolizes a fe-

King of the jungle: Not another deadbeat dad

male. Males in promiscuous troops such as these copulate with as many females as they can. Thus the possibility of a nonpromiscuous female monopolizing the investment of any male by reducing her copulation rate and his paternity uncertainty is opposed by the male strategy of promiscuity.

To support the argument that orgasm is a motivational device designed to increase a female's tendency to copulate, orgasm should be most developed in species which copulate frequently and females in such species should initiate sexual activity. In species with very low female copulation rates and little variation in male quality or both, we expect less solicitation and less orgasm. Frequent sex was not expected in and, indeed, is not a feature of the lives of tamarins, marmosets, titis or night monkeys, all of which are known for monogamy and high male parental care. It is in promiscuous monkeys, such as various macaques and chimpanzees, that orgasmic behavior was first documented and shown to be most comparable to that of human females. But this correlation is weak at best and would be hard to establish, since it is difficult to quantify female orgasm in other species. The report of female orgasmic behavior in gorillas argues against orgasm and increased copulation rate as being ways to test males.

These are all complicated scenarios, and none of them seems to capture any generalizations about female orgasm. A more mechanical explanation may turn out to be more general. Experimenters have found that orgasm in human females results in a sharp change in air pressure in the uterus. Before orgasm, the air pressure in the uterus is positive, but at orgasm, it reverses, and suction is created. This would have the effect of drawing sperm up into the uterus and increasing the chance of fertilization. In other words, orgasm would be a form of female mate choice, allowing the female to exert some control over who fertilized her eggs. Since the uterus can be a formidable barrier to sperm, orgasm under female control could be an effective device for enabling her to decide the fate of the male ejaculate according to her interests.

So it may all come down to something as prosaic and simple as suction. But that does not diminish the true evolutionary import of the female orgasm: In orgasm, females may take a proximate pleasure in achieving an ultimately adaptive goal.

CHAPTER 12

Smelling

"The smell and taste of things remain poised for a long time, like souls, ready to remind us, waiting and hoping for their moment, amid the ruins of all the rest; and bear in the tiny and almost impalpable drop of their essence, the vast structure of recollection."

Marcel Proust
Swann's Way

A friend smells smells that are lost on me. Just yesterday, she sniffed out a tropical tree some hundred yards away, exclaiming about the power of the scent. I smelled nothing until I pushed one of the deep, creamy white blossoms flush with my nose, and then I detected only a mild musky odor. It is the same when we go mushrooming; she is always the first to realize that stinkhorns are about. Stinkhorns are dramatic mushrooms with names like *Phallus impudicus*. They stand erect from the soil in a form so true to their name it is scarcely credible and produce an aroma that brings in sticky-footed flies to disperse their spores. I usually see them as soon as I smell them. Our difference in olfactory acumen has nothing to do with, say, the difference in the size of our noses. It is simply male and female in its division.

Women can smell some scents that men only dimly perceive. It's no accident that the tree I mentioned registered on my friend's nasal nerve endings and not on mine. It is pollinated by bats, and like most bat flowers, it has a musky scent. Musk is a sexual scent, a fact that may or may not be appreciated by perfume purchasers. Eugene Marais, the visionary South African biologist who wrote *The Soul of the White Ant* and *The Soul of the Ape*, commented: "Ask a young woman why she uses the heavenly perfumes the chemist of our day has learnt to concoct in such exquisite perfection. Her answer will be misleading, because she does not know the subconscious reason.... She would be embarrassed if she learnt the basis of all her perfumes were the sexual secretions of several kinds of cats, a deer and (the most expensive of all) the rudimentary sexual material secreted by a... whale."

Our sensitivity to musky smells is affected by our sex. One famous experiment comparing the olfactory sensitivities of males and females to various chemicals showed that women are 100 times more sensitive than men to exalotide, a compound similar to male sexual musk. A woman's sensitivity to this musk peaks about the time of her ovulation. It can be completely extinguished if a woman has her ovaries removed, and it can be restored by injections of the female sex hormone estrogen. In the days when most hogs were home-raised and many male pigs were not castrated as promptly as they are in today's efficient factories, "boar taint" was a common complaint of women diners. Women are far more sensitive than men to the musky taint of the meat of a mature male pig. This sexual dimorphism in sensitivity suggests that odor communication may be part of human sexual interactions, but little attention has been paid to the topic. Nevertheless, in other species of mammals, smell is a leading form of sexual communication.

In some mammals, a whiff of a single chemical compound is enough to induce sudden sexual interest in male or female. If a female dog, one not in estrus, is dabbed with methyl p-hydroxybenzoate, all male dogs in the vicinity are aroused. Mammals use smells and tastes for signaling their sexual status and for reading and tasting olfactory information about other animals. Some secretions are extremely complex. The pheromones (chemical signals) put out by beavers contain 50 different molecules, the proportions of which vary from individual to individual. Theoretically, this is enough to give every beaver on earth its own unique chemical signature and to provide a vast array of information. In fact, beavers do recognize the scent of foreign beavers. The standard trapping technique is to smear some beaver-gland secretion on a stick near a trap set along the shore or dam. As soon as the resident beavers detect the smell of a strange beaver, they become agitated, slapping their tails and searching for the trespasser, the resident male looking to expel males and the resident female to expel females.

Many large grazing ungulates such as camels, horses and deer smell and taste urine to judge their mates. Males may harass and kick at females until they urinate, or females may do so voluntarily. When they do, the male will swallow the urine and draw back his lips and nostrils into a characteristic grimace (flehmen), a posture thought to increase sensitivity, similar to a wine taster swirling wine in the mouth to read its flavor better. Urine-tasting enables a male to assess the reproductive condition of females.

Males may also give off perfume. Billy goats, for example, spray urine and even ejaculate into their beards, a practice that contributes to their goaty odor and must convey information on their sexual readiness. White-tailed deer and elk bucks urinate and ejaculate on the ground and wallow in it, presumably for similar reasons.

Male porcupines feature their urine while courting females. Here is one observer's account of the proceedings that so many have speculated about: "The male usually coaxes the female to the ground, where he will rear on hind legs and tail while emitting low vocal 'grunts.' He then proceeds to spray the female with bursts of urine from a rapidly erecting penis, and after wrestling chases, vocalization and more urine showers, coitus is effected. It is performed, as in most mammals, with the male taking the active role and, contrary to folklore, mounting from the rear. The receptive female elevates her hindquarters and arches her tail over her back, providing the male with a platform for his forepaws or chest. The male then permits his forelegs to hang free. Coital contact is brief, with violent ejaculation, and afterward, the male drops back to groom and clean. Further matings may ensue until one of the pair climbs a tree and ends the contact by hostile screaming and lunging."

Rabbits and hares also use urine in courtship. A courting male snowshoe hare entices a female with great leaps. While he is airborne, he showers the female with urine. Presumably, this impresses the female with his hormonal state. As is the case with Olympic athletes, a sample of urine provides a means to estimate the amount of steroid sex hormones such as testosterone the urinator possesses.

Humans have begun to put these discoveries to commercial use. My NASCO farm and ranch catalog offers aerosol spray cans of a product called Boar Mate. Farmers can spray their sows into lordosis, the posture and attitude needed for boars to be able to mate them. The compound in the can is androsterone, or androsterol. Androsterone is found in pig saliva and is used for sexual communication between male and female swine and possibly between humans. It was first found in human urine some 30 years ago. More recently, it has been found in the armpits of humans.

Of all mammals, humans have the most highly developed armpits. The axillary, or armpit, scent-gland complex is found only in humans, gorillas and chimpanzees, animals that spend much of their time in an upright posture. This posture may favor the armpit as a site for chemical transmission. The tuft of wiry hair bristling from the armpit is much like the hair pencils that various moths and butterflies extrude from their hind ends when they broadcast their mating pheromones. The large surface area of the hairs acts as a radiator of chemical molecules, increasing the rate at which they float off into the air.

The sweat glands in our armpits do not excrete androsterone or other musky-smelling compounds directly. Pure sweat is virtually odorless and remains so for at least two weeks, but sweat exposed to the armpits' rich community of bacteria develops a strong odor within six hours. A certain taxonomic group of bacteria is responsible for producing the smell. It is the relative abundance of these bacteria (aerobic diphtheroids) that makes some people smellier than others. The armpit is highly favorable for these bacteria, which are able to rebound after valiant efforts to expunge them. This suggests that we have been designed to encourage their presence because we have derived some benefit from them—that is, the smell they produce—in the past.

One of the compounds the bacteria operate on is cholesterol. It is excreted in the armpit at times of excitement and stress; an injection of Adrenalin, for instance, will boost the secretion. Cholesterol is one of the precursors the body uses to make sex hormones such as progesterone and testosterone as well as the androsterone-type compounds that the bacteria convert into androsterone.

Humans are extremely sensitive to the odor of androsterone and are able to detect it at a level of 0.001 to 0.005 nanograms per microliter, a measurement that is not readily explained in human terms. Suffice it to say that the human threshold—what our noses can detect—is beyond the measuring ability of sophisticated chemical-analysis machines such as gas-chromatographic and mass-spectrographic analyzers worth tens of thousands of dollars. Whether humans make use of their sensitivity to androsterone is under dispute.

Shakespeare's King Lear says: "Give me an ounce of civet, good apothecary, / Sweeten my imagination." And some cosmetic firms have taken to selling colognes and cosmetics containing androsterone. But quantifying the human effect is difficult. In one study, researchers sprayed theater seats with androsterone to find out whether the sprayed seats were more attractive to women than to men. In another study, men and women were shown photographs of people and asked to make judgments about how attractive they were. The people who were rating the photographs were not told they were involved in an odor experiment, but some photographs were presented when there was androsterone in the testing room. Both men and women rated the photographs as sexually more attractive when they were exposed to androsterone during the viewing. A similar study showed that androsterone could favorably influence the impression a woman interviewer received when interviewing a male job applicant.

Testing for the effects of androsterone on humans is difficult because context so greatly influences human reactions and impressions. One test had students reading an erotic piece of literature while breathing through a mask containing either androsterone or a control rosewater solution. No arousal was detected in either group because of the artificiality of the testing conditions. Similarly, the prestigious journal *Science* published a study entitled "Changes in the Intensity and Pleasantness of Human Vaginal Odors During the Menstrual Cycle." It reported that the intensity of vaginal odor declined as ovulation approached and that both male and female subjects found it less pleasant at that time. But, the investigators concluded, the data "do not support the notion that such odors are particularly attractive to humans in an in vitro test situation." The last phrase means that sniffing vaginal swabs while sitting in an air-conditioned room is not a particularly enjoyable task.

Human vaginal odor involves at least 30 compounds derived from "(i) vulvar vaginal secretions from sebaceous, sweat and Bartholin's and Skene's glands, (ii) mucus from the cervix, (iii) endometrial and oviductal glands, (iv) transudate through the vaginal walls and (v) exfoliated cells of the vaginal mucosa. Many or all of these components may produce and/or provide substrates for bacterial production of volatiles." But such a strong and complex odor taken out of its natural context is no more likely to be perceived as pleasant than is a whiff of Limburger cheese or anchovy paste, fragrances many find extremely appetizing in the right situation. So it is impossible to say conclusively whether any data indicate a pheromonal attractive or excitatory role for such odors. The literature of human emotions certainly contains much anecdotal evidence in favor of the idea.

While some companies continue to sell armpit deodorants by the ton, attempts by cosmetic corporations to patent and market human pheromones continue. The fatty acids in vaginal secretions, known as copulins, have reportedly been patented but not commercially marketed. This may not be so much because of their somewhat fishy fragrance but because they may attract the attentions of monkeys, billy goats and other inappropriate mammalian suitors. Whether or not it will do anything to human males is unproven.

There is a remarkable twist to the story of androsterone. As we have said, hogs of both sexes exude androsterone in their saliva. The same compound is given off by truffles, perhaps to attract vertebrates that will eat them and disperse their spores. It is undoubtedly why hogs are such excellent truffle finders and perhaps why people are willing to pay outrageous sums for a whiff of the dank truffle odor. But before you rush off to buy a can of Boar Mate for your own sexual experiments, you might consider the fact that some of the women in the Royal Shakespeare Company, the theatrical group that participated in the study of the effect of androsterone on the choice of seats, found their menstrual cycles disrupted. And anyone who keeps company with swine ought to be especially cautious.

Another olfactory phenomenon ranks as a sexual curiosity and points out how we may be totally unconscious of the effect of odor on our physiology and behavior. Martha McClintock, a specialist in olfactory communication and the reproductive biology of mammals, was the first to quantify and record menstrual synchrony among cohabiting women. She found that women living in her college dormitory tended to menstruate at roughly the same time. Women arrived at the dormitory out of phase with the others, but as the year progressed, they all began menstruating at the same time of the month, settling gradually into phase after four cycles. This ovulatory synchrony is achieved by olfactory signals contained in human sweat that bring the hormonal cycles of individuals into phase. This has since been independently confirmed, but its significance remains obscure.

The time of menstruation is only an indicator of a more important phenomenon—ovulatory synchrony. Some biologists have suggested that ovulatory synchrony is an adaptation resulting from human polygyny. Nancy Burley suggests that ovulation synchronized because it enabled the polygynous male to predict when his wives were ovulating. Having several wives out of phase would have made things too complex for him, and it might be difficult for him to fertilize all his wives. Thus by synchronizing the rate of fertility, the number of impregnations could be raised. Women who synchronized would then reproductively outcompete women who did not synchronize. There are several problems with this idea. First, the number of highly polygynous marriages in human society is far lower than the number of monogamous ones. Second, there is no evidence for the idea that women find it difficult to get pregnant. In fact, the reverse is true; they expend much effort trying to avoid becoming pregnant too frequently.

Richard Kiltie, a biologist who believes synchronization is an epiphenomenon, reasons that a female who was not part of the synchronized cohort would be the only one receptive when the other wives or cohabiting women were menstruating and so she would be most likely of all to get pregnant. In and of itself, this should select against the phenomenon. Kiltie also notes that cohabiting males show synchronicity in monthly body-temperature cycles and in the excretion of various hormones, and males may synchronize these with the menstrual cycles of their wives. In his opinion, synchronicity may be nothing more than some old evolutionary vestige in humans, an ancient tendency to synchronize bodily cycles.

An alternative explanation is that menstrual synchrony is a female strategy to minimize the harmful effects of male-male combat and infanticide. Male aggression may cause significant mortality to the female's offspring. If all females are reproductively active at the same time, infant mortality would be reduced. On the other hand, synchrony might be a result of female-female competition for access to dominant males and reproductive rights. There is evidence that female baboons, wolves, tamarin monkeys and rodents inhibit each other's ovulation and reproduction.

Synchrony could conceivably be a strategy adaptive to both the dominated and the dominant. Dominant females would be better able to monopolize the dominant male and reproductive resources such as den sites and food if everyone was in estrus at the same time. Work with rodents supports this idea—it is subdominant females that bring their cycles into phase with the dominant. Thus ovulatory synchrony may be olfactory coercion.

Olfactory coercion is a specialty of rats, mice and many other social rodents. As social mammals, they may give us some insights into the forces—even if they are now irrelevant—that were responsible for evolving menstrual synchrony. Especially sensitive to the scent signatures in urine, rodents adjust their reproductive state according to what they read. Juvenile female house mice exposed to the urine of a mature male will reach estrus 20 days before unexposed females. If an adult female mouse is exposed to the urine of a male, her uterus weight begins to shoot up within 20 minutes and her reproductive hormones are triggered into action. In prairie voles, this effect is so pronounced that the weight of the female uterus doubles in two days. On the other hand, if more than one female is present, the effect is reduced. Even when male urine is present, female urine will inhibit estrus in another female by seven days, and if there is no male around, the inhibition will last as long as four weeks. The effect also operates on males. Urine from females accelerates male maturation, while urine from other males inhibits it.

McClintock has shown that urine from female Norway rats contains two different airborne cues that tend to cause menstrual synchrony within the social group. One cue shortens the cycle and enhances the probability of coming into estrus, while the other lengthens the cycle and suppresses estrus. It is possible that dominant and subdominant females differ in the proportion of the two chemical cues they produce. Subdominants may signal their submissiveness as a payment for freedom from attack by producing the accelerating pheromone, while dominant females ought to produce the suppressing compound.

And of course, a male's urine ought to attempt to countermand these signals chemically. His urine should be designed to bring as many females as possible into estrus as quickly as possible. It is known that male urine can influence female reproductive physiology to the female's detriment by means of the famous Bruce effect, named for the late physiologist Hilde Bruce. She discovered that if a pregnant mouse is exposed to the urine of a strange male, she either fails to implant the embryos or spontaneously aborts implanted embryos. This is another form of male infanticide, in which a male destroys the offspring fathered by another male and attempts to bring the female back into estrus, a ready receptacle for his own genes. At first, this seems totally maladaptive from the female's point of view. Yet it may be a way for her to minimize her loss of investment. If the new male owner of the territory will destroy her offspring, then she can cut her losses by aborting immediately. Nevertheless, she still suffers as a result of the infanticidal male strategy.

Sympathizers of female rodents will be gladdened by the way the olfactory acumen of female brown lemmings affects the males' behavior. Females can smell success. Male brown lemmings fight for access to females, and the outcome of their battles influences their attractiveness. When the female is given a choice between a winner and a loser—both still capable of mating—she chooses the winner, and she does so by using her nose. The female can distinguish between winners and losers within five minutes after a fight and up to an hour or even a day later. This is probably because the outcome of the battle influences the hormonal physiology of the combatants.

This study has an interesting parallel with a study of the Harvard wrestling team. A survey of blood testosterone levels found that males who won their matches had a much greater rise in testosterone levels than losers did. It is easy to see how this might be adaptive. Testosterone—the male sex hormone produced by the testicles—increases male aggression. Leonardo da Vinci pointed this out in his notebooks: "The testicles increase the animosity and ferocity of animals. The principle is clearly illustrated in the case of castrated animals, for one sees the bull, the boar, the ram and the cock, very fierce animals, which having been deprived of their testicles remain very cowardly; so one sees a ram drive before it a flock of wethers, a cock put to flight a number of capons." But if a male's testosterone level rises above his fighting ability and he engages in battles in which he will be outmatched and injured, then he will suffer reduced fitness. Thus losers of a contest ought to decrease their combativeness, and winners ought to raise it. Success in combat should set the testosterone-aggressiveness thermostat. And if fighting ability is a valuable heritable trait, then females ought to select for males of proven ability.

Aggressiveness is important for lemmings (and probably for students enrolled in the Harvard M.B.A. program). Lemmings periodically undergo intense crowding as a result of exploding populations, and they must frequently fight for territories. Aggressiveness is a heritable characteristic of many rodent species, so there is value for females in choosing winners over losers.

It would seem valuable to be able to know things by olfactory means, but textbooks tell us that we have abandoned it; our sense of smell has become superfluous as our eyes have grown keen. Yet I think it is too soon to abandon the idea that the human sense of smell can influence our behaviors. Smell may still be bound strongly to unconscious imprinting and memory associations. We have a good

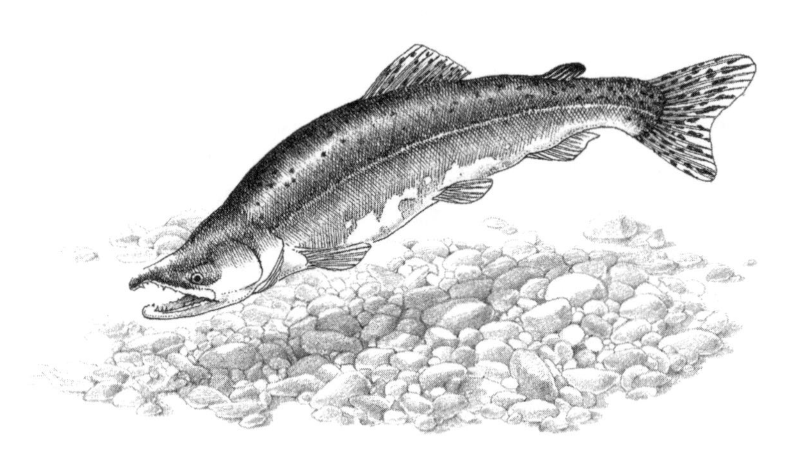

Male salmon: The nose knows the way home

example in the way one man, Arthur Hasler, used his experience of olfactory imprinting to gain insight into the biology of salmon homing. In *The Compleat Angler*, Izaak Walton noted that the practice of tagging salmon with a piece of string "has inclined many to think that every salmon usually returns to the same river in which it was bred." Yet how could a salmon swimming in the ocean ever find its way up a forking labyrinth of tributaries and back to a tiny mountain streambed? The mystery remained unanswered for three centuries.

Hasler argued correctly that salmon sniff their way home. Here is his account of how he came on the idea: "As I hiked along a mountain trail in the Wasatch Range of the Rocky Mountains, where I grew up, my reflections about the migratory behavior of salmon were soon interrupted by wonderful scents that I had not smelt since I was a boy. Climbing up toward the alpine zone on the eastern slope of Mount Timpanogos, I had approached a waterfall which was completely obstructed from view by a cliff; yet when a cool breeze bearing the fragrances of mosses and columbine swept around the rocky abutment, the details of this waterfall and its setting on the face of the mountain suddenly leapt into my mind's eye. In fact, so impressive was this odor that it suddenly evoked a flood of memories of boyhood chums and deeds long since vanished from conscious memory.

"The association was so strong that I immediately applied it to the problem of salmon homing. The connection caused me to formulate the hypothesis that each stream contains a particular bouquet of fragrances to which salmon become imprinted before immigrating to the ocean and which they subsequently use as a cue for identifying their natal tributary on their return from the sea."

Our sense of smell is still strong enough to haunt and beguile us. Who has not had some brief fragrance—the smell of a flower or a summer rain—unleash memories and images of scenes and emotions long past that come surging back in clarity and detail. What memories may be filed under and conjured up so richly by the scent of orange blossoms, and what others wait for the scent of nutmeg or a thousand other redolences for their release? This link between smell and memory may now be mere atavism. But from the power of such associations, we may gain an inkling of how it used to be.

CHAPTER 13

Sex Change

"It has now been ascertained that at a very early embryonic period, both sexes possess true male and female glands."

Charles Darwin
The Descent of Man

Female shrimp, be they steamed or stir-fried, make much better eating than male shrimp do. Females are generally larger than males, allowing a fuller gastronomic appreciation of the resilient crunch of crustacean meat. Female jumbos, however, do not attain their large size or command a high price simply by being born female. In fact, most females start life as piddling, low-priced males.

Commercial shrimp, usually members of the genus *Pandalus*, change sex. We may consider this strange. After all, we are the only mammals that attempt it and then only in a cosmetic sense. But for many plants and animals, sex change is a normal part of growing up. Our own condition of inflexible gender makes it easy to understand why other groups do not change sex. For sexually dimorphic species with internal fertilization and all the complicated constructions and behaviors it requires, the costs of switching from one system to another must be considerable. I have never priced the cost of human sex change, but I would bet it is expensive. The question, then, is why, for some organisms, the benefits exceed the costs of conversion.

Some cases of sex change seem to be associated with life histories in which males and females use grossly different ecological resources. As an extreme example, we have the life history of a marine sowbug, *Danalia curvata*, a parasitic isopod that feeds on other ocean dwellers. *Danalia* begins life as a planktonic larva; it is sexless, with neither testes nor ovaries. It looks like a tiny, living grappling hook, a millimeter-long tangle of legs, hooks and claws devoted to snagging a host. Unless a *Danalia* can latch onto a free-swimming copepod, another zooplanktonic organism, all is lost. *Danalia* that achieve their attachment transform into males, and as their testes develop, the legs and hooks disappear. I suspect they are in part self-digested and reconstituted as sperm, for mating is the only task of the male. When it finds itself in the vicinity of a female *Danalia*, a perception probably based on smell, the male empties his testes and becomes a female.

The female, too, is a parasite, but a more ambitious one. Her host is a barnacle, *Sacculina*, which looks nothing like the little white calcareous cones on rocky shores or the goosenecks on floating driftwood. *Sacculina* is itself a parasite, an orange sac found on the underside of crabs. The sac has a branching stalk that grows its way through the crab's body like a root in the ground, sponging up nutrients. The female *Danalia* does likewise unto the barnacle. When the still male *Danalia* contacts an appropriate barnacle, he, too, turns baggish. His testes dissolve, and he grows a giant, invasive proboscis that worms its way into the body of a *Sacculina* and begins to feed. Ovaries then form, and we may say the male is now female. She grows large, 13 times larger than her formerly masculine self.

The association between size and sex is not coincidence. The changes go together in other species, an observation which led Michael Ghiselin to suggest that sex change can be explained by the size-advantage model: Sequential hermaphroditism—that is, the change from one sex to another—will be favored if an individual's performance as a male or female depends strongly on size in one sex or the other. In other words, if a female does much better when she is large than a male does, selection will favor an individual that changes sex. And it will favor an individual that makes the switch at the right size, a point which maximizes the sum total of the individual's success as a male plus its success as a female.

We can see why *Danalia* should be female when it has the opportunity to grow large by feeding on a fat barnacle-crab dinner. She is an egg machine and nothing more: Her success is related simply to the total number of eggs she can broadcast into the ocean. A male does not need to be large. Sperm are tiny, and a small male can make all the sperm needed to fertilize the eggs of a female. Males do not fight or guard females, so there is no value in being large. In fact, being small may increase the number of copepod hosts a male can use and thus increase the chances that he and his brethren are dispersed to find females.

The pattern of sex change from male to female, known as protandrous sequential hermaphroditism, is common when there is strong fecundity-size selection on females and little reward for being a large male. It is especially common in marine organisms that have no internal fertilization and broadcast huge numbers of eggs or larvae into the open ocean. Oysters are another example. In a bed of 1-year-old oysters, 70 percent of the individuals will be male. By the next year, after they have grown, the sex ratio is half and half. The following year and afterward, they are mostly females.

Oysters have a simple social life. Males and females—and, for that matter, males and males and females and females—interact little. Their life is just an endless swallow of seawater to be filtered and converted into more oyster. But in other marine organisms, more complex social interactions often produce the reverse sex change, from female to male. This pattern is common in coral-reef fish families, including parrotfish, wrasse, goby, grouper, damselfish and angelfish.

When I first read that sex change was highly developed among reef fish, I thought the correlation spurious. Coral reefs are exceedingly pleasant places to work, and they are far more accessible than other habitats such as the Amazonian drainage or the deep sea; I thought researchers were just looking for an easy habitat to study, but there seems to be more to it than that. Reef is a three-dimensional and exceptionally clear visual habitat. It is predictable in its tides and its lunar and daily cycles, and it is real estate carved up into stable territories in a way that the shifting floor of a muddy river could never be.

The fact that social interactions are of paramount importance in the relations among reef fish is also demonstrated by the brilliant colors of both sexes. These are not warning colors that advertise poisonous flesh. Many of the most gorgeous and conspicuous reef fish are highly palatable. It seems that the colors are probably used in social signaling and the establishment of dominance relationships.

In a typical protogynous system—one in which a female becomes a male—a territorial male guards a piece of reef as a means of obtaining exclusive access to a harem of resident females or females that use the territory in some way. The cleaner wrasse, *Labroides dimidiatus*, affords a typical example. This species is found on Australia's Great Barrier Reef, where it gets its food at traditional spots known as cleaning stations. Other reef fish visit these stations, and the wrasse removes and eats the copepods and other external parasites they carry. Each station is lorded over by a dominant male and about half a dozen females. He mates with each female once a day. A simple experiment with this social system had a remarkable result. When the male harem master is removed, he is not replaced by some surplus waiting male from outside the system—there are no males outside the territory. Instead, the largest female in the harem promptly changes sex. She turns into a male within hours, and within a few days, she is a producing, viable male, mating with her own set of females.

This phenomenon is again best explained in terms of the relative-size advantage. A small male has virtually no chance of breeding because of dominance and territorial relations, but a small female can always breed. Thus natural selection favors individuals that start off female and stay female until they are large enough to assume control of the territory as a male.

It follows that reef fish ought to be assessing, albeit hormonally, their relative size. This has been shown for the Hawaiian saddleback wrasse, *Thalassoma duperrey*. In experiments, individuals changed size according to how large or small their neighbors were. Small companions stimulated an individual to grow, and large ones inhibited it.

This observation was fine-tuned under natural conditions. An experiment with sea bass, *Anthias squamipinnis*, in natural groups on a coral reef showed that male removals led to incredibly precise changes in female social groups. This is notable because female groups are large, containing an average of 61 and up to as many as 294 individuals. Experimenters removed 58 males from 11 groups. Remarkably, 57 females changed sex, and even more remarkably, the sex changes were serial—the largest female in a group changed first, then the next largest, and so on.

Sex change in these systems seems to involve individual assessments of local conditions. The size advantage will depend on which individuals are competing, how large they are and how many there are. This is borne out by Robert Warner's extensive work on the blue-headed wrasse, *Thalassoma bifasciatum*, a native of Caribbean reefs. These fish are common—anyone who has snorkeled in the Caribbean will have seen them—and they are convenient, since one can distinguish females, males born as males and males that were once females by their turquoise, blue and white patterns. Better still, the reefs they occupy vary greatly in size, all of which means that there is opportunity to assess how the male size advantage changes ecologically.

Warner found that on small reefs, the large-male-size advantage was great. Big males could defend territories effectively and gain virtually exclusive sexual access to up to 150 females daily. But as the reefs got larger, female spawning sites became less and less defensible. Females aggregate to mate and spawn at the end of a reef; on large, populous reefs, the aggregation is too huge for any one wrasse to dominate and control. Under these conditions, small male individuals can also fertilize eggs. Warner found a neat correlation showing that the size advantage depends on the defensibility of females. On small reefs, where males can defend females and spawning sites against other males, only 20 percent of the males were born as males. The rest started as small females and turned male only when large. But on the large reefs, where small males can fare well, as many as half the fish are born male.

If males are monogamous and invest parental care in their young, large size may be of little value to them. In fact, if they invest heavily, it may be that the females compete for male parental investment, so the large-size advantage will go to females. This is the situation found in clownfish in the genus *Amphiprion*. These fish live in mated pairs in association with sea anemones. The male lives with one mature female and provides egg-guarding services for her. Females are larger than males, presumably because of the precedence of fecundity selection. If a mature female is removed, her mate changes sex and becomes the mate of one of the smaller males hanging around the territory.

It is necessary to confront a question: Why change sex at all? Why not simply add a sex and retain the flexibility of being both male and female? Simultaneous hermaphroditism is possible, and it does have decided advantages. Michael Ghiselin has argued that if an organism is hermaphroditic, the costs of mate-finding ought to be less. He predicts—and has some support for the idea—that being both male and female is favored in a rare organism. The price for this is the cost of producing and maintaining two sexual systems.

Sperm may be relatively cheap, but the apparatus needed to proffer it is not. In the slipper limpet, *Crepidula fornicata*, males form stacks on top of females. Slipper limpets begin as sexless larvae. A female is simply the individual lucky enough to get into place first. Settled on a bare area of rocks with no other females, she is free to grow as an ever-enlarging egg machine, controlling with pheromones the sex of the individuals that settle on her. Males will remain males for 18 months when exposed to a female but for less than 6 months if she leaves. A stack of these limpets may grow as large as 20 individuals, which places a constraint on male success. To reach females, a male must invest in a penis that is much longer than his body. It is an erectile organ that doubles in size. Erection in itself implies a cost. It suggests that a penis in its enlarged state is a bulky, cumbersome appendage that gets in the way of other tasks, so when not in use, it is collapsed and, ideally, tucked away in a protected spot. The penis of the slipper limpet is also coated with lumpy structures apparently designed to lodge it firmly in the female so that other males will not easily displace it. That all this is costly is suggested by the fact that the penis waxes and wanes in size according to the breeding season.

A simultaneous hermaphrodite bears more than just physical and physiological costs. There are behavioral costs as well. Simultaneous hermaphrodites such as earthworms have no complicated penis to pay for, but they may pay a price for tactical complexity. We have every reason to expect subtlety in the sex lives of earthworms. Since each worm may act as both male and female or just male or just female when it meets another male-female or male or female, there is potential for conflicts of interest beyond those attending a simple male-female interaction.

The most obvious possibility is cheating. Observers of the turbellarian worm *Stenostomum oesophagium* report that these worms are simultaneous hermaphrodites which normally cross-fertilize but that some individuals will inseminate others while slyly avoiding becoming impregnated themselves. Other reports of this sort of behavior among turbellarians suggest that individuals will impregnate promiscuously but allow themselves to be inseminated only selectively. Given that sperm are cheaper than eggs, a hermaphrodite may attempt to invest more in sperm transfer than in egg production. Each successful sperm will contribute as much as a fertilized egg and be cheaper. The cheater would attempt to fertilize all the eggs of his partner but offer none in return. This strategy might easily spread through a population until it stabilized—half the population would be sperm specialists and the other half egg specialists—or until the cheating was contained by some behavioral strategy.

Evolutionary biologist Eric Fisher has described one such behavioral strategy used by hamlets, *Hypoplectrus*, small sea bass from the Caribbean that are simultaneous hermaphrodites. Hamlets trade eggs.

They mate in pairs, and both partners have eggs and sperm to offer. To prevent cheating, neither partner releases all its eggs at once for the other to fertilize. Instead, they reciprocate—one partner releases a few eggs for fertilization and then the other—for up to four bouts, until their eggs are exhausted.

Unbeatable tactics may exist. Slugs in the genus *Ariolimax* are, like most slugs, simultaneous hermaphrodites. Normally, they mate as reciprocals, each partner inseminating and being inseminated. *Ariolimax*, however, often attempt and achieve appophalation; that is, one manages to gnaw off the other's penis. The deprived slug cannot regrow his penis and is now obligated to be a female and is forced to offer eggs. The penises of slugs are huge, so dispatching a penis in this way may simply be a gastronomic endeavor, although a risky sort of meal to seek. However, it may be that the aggressor can raise his reproductive success by increasing the local density of females. No evidence for this exists, but it is known that slugs are sedentary and territorial, so the idea has potential. If it is possible, it would again destabilize the simultaneous strategy.

It is interesting that the largest class of simultaneous hermaphrodites is flowering plants in which the interaction between individuals is indirect and mediated by pollinating animals or the wind. Not only does this give plants less scope for cheating, but since male and female parts occur together and pay the same price—usually nectar to attract a dispersal agent—there is no differential size advantage in being male or female. In fact, a flower may be most successful if it employs the same pollinator to both spread and accept genes. The sum of the reproductive success of the two genders may be greater than if it produced all-male or all-female gamete dispensers (flowers).

There are true sex-changing plants, and they tend to be plants in which there is a size advantage. The best-studied of these is the jack-in-the-pulpit, *Arisaema triphyllum*, a common plant of the eastern United States and southern Canada that blooms in rich, moist deciduous forests in early summer. The size advantage explains why this plant changes sex. Small plants with a single leaf are almost invariably male. They cannot pay the price of pregnancy. The crimson fruits of the jack-in-the-pulpit are large, waxy berries with a big seed that are dispersed by birds. A female, which may be 20 years old and have large food reserves stored in the roots, can mature dozens of fruits. But for a small male to mature even half a dozen fruits would consume al-

Jack-in-the-pulpit: Nature's switch-hitter

most a third of the plant's resources. No perennial plant can invest this heavily and stay perennial. That degree of commitment is found only in annuals, which die after reproducing. When the jack-in-the-pulpit grows large, it turns its size to advantage by producing seeds, which cost more than pollen but have a far greater chance of growing into descendants. Similarly, a female that has a bad year caused by, say, the attack of a herbivore will revert to being male the next year.

The size-advantage model, then, makes sense of sequential hermaphroditism. But as one reviewer of the topic lamented, the theory has not proved capable of predicting why some species do not change sex. It does not tell us why jack-in-the-pulpits change sex while another flower, say, a marsh marigold or some close relative such as a calla lily does not. It is a model that makes sense only for specific cases. I do not think this is a fault of the size-advantage model. That it explains the life history of a single species is accomplishment enough in ecology and evolution. So little has been explained. And although it is specific, it contains a message that is as general as they come: Males and females need different means to accomplish the goals of their genders.

CHAPTER 14

Incest and Outcest

"Almost everyone carries the equivalent of more than one lethal gene in the recessive heterozygous condition."

L.L. Cavalli-Sforza and W.F. Bodmer
The Genetics of Human Races

The Greenland Inuit have a legend about the sun and the moon. To amuse themselves in the dark of winter, the inhabitants of an igloo would set their seal-oil lamps outside. In the darkened igloo, they could then make love anonymously. No one could be exactly sure who his or her partner was. A girl who joined this game every evening began to suspect that her brother was making love to her. One night, she coated her hands with soot and smeared her lover's face. She then ran outside to get a torch. Her lover followed, and by the light of her torch, she could see that he was indeed her brother. She took her torch and began to run, and he chased her, carrying his own torch. They ran on and on, faster and faster, until they began to rise, spiraling into the sky, "And the girl, warm and shining, / Became the sun / And her brother, dark and cold, / Became the moon."

Biologists should see a genetic moral in the story. The woman who ran from incest is shown in a favorable light—the sun, giver of light and warmth, while the incestuous brother is diminished, dark and cold. But why should the Inuit abhor incest? Some stringent geographical factors could favor a tolerance for incest in the high Arctic.

At the top of our planet, people used to live a harsh, isolated life. Inuit villages could survive only at special sites where hunting conditions were suitable. Often, communities were separated by hundreds of miles, distances made even vaster and more formidable by the cold and by winters during which the night darkness lasts for months. For a human in search of a mate, such obstacles might seem to favor inbreeding. It would be far easier to find someone within the group, even if he or she were closely related. And yet by their legends and taboos and their mate-choice behavior, the Inuit show no tendency to inbreed. Marriages between even sixth cousins were once prohibited before acculturation took place. The Inuit, like humans the world over, had rules about mate choice.

The fundamental rule in human mate choice is this: Incest is taboo. The difference between classical anthropological explanations of human-incest taboos and a biological explanation comes down to two perspectives—seeing selection in terms of individuals and seeing selection in terms of groups. No one disputes that humans abhor incest. With a few minor exceptions to be discussed later, every culture prohibits primary incest—that is, brother-sister or offspring-parent matings. Most forbid marriages between cousins. But the simple biological explanation for this has not been universally accepted.

The biological explanation of incest avoidance is simply that incest is selected against by a phenomenon known as "inbreeding depression," in which the offspring produced by incestuous matings are less fit than those from outbred matings. This explanation is easy to follow if one is willing to wade through a little genetic jargon.

Inbreeding depression follows inevitably from the mechanics of our genetic system. Organisms such as ourselves are diploid, meaning we contain two sets of chromosomes, two sets of genes arranged and organized into particular sequences and combinations. Because we have two sets, every gene within an individual can exist in two forms. These alternative forms are called alleles. For example, the gene that controls brown eyes can have several allelic forms. At any given gene, an individual can carry two alleles that can be identical or different. If they are identical, the gene is said to be homozygous; if they are different, it is heterozygous.

Some allelic forms are nonfunctional, or harmless; others may be dangerous—coding for a protein that causes death or reducing the vigor of the carrier, for example. If these deleterious effects are expressed directly, natural selection will quickly remove them from the population. But some defective alleles are not expressed if they occur with a functional allele. Consider, for example, a gene that controls the production of skin pigment. An individual might contain a defective albino allele that would result in no protective pigment being produced, but if the other allele was functional, it would code for pigment production, and the individual would appear normal. Thus functional alleles may "dominate" and mask the effect of an unexpressed—"recessive"—and deleterious allele, lowering the rate at which deleterious recessives are exposed to the culling action of natural selection.

The deleterious recessives are part of what geneticists call the "genetic load" of a population. Every population carries a significant number of deleterious alleles, and every individual usually carries deleterious alleles that can lower his or her fitness. The amount of the genetic load is measured in "lethal equivalents." For example, if an allele lowers an individual's fitness by 20 percent and an individual has five deleterious alleles, then that individual carries one lethal equivalent (five times 20 percent). Human populations are estimated to carry an average of two to five lethal equivalents per person. Similar estimates have been made for fruit flies, flour beetles and pine trees.

Since most of these deleterious alleles exist in an individual as unexpressed recessives masked by functional, dominant alleles, an individual rarely suffers from their effects. Deleterious effects are usually expressed only when an individual has two deleterious alleles of the same gene, in which case their defective nature is exposed.

For individuals who breed with nonrelatives, the probability that both will contribute the same set of deleterious alleles to their offspring is very low. Each individual contains tens of thousands of genes, and there may be many allelic forms of each gene, so the chance of an offspring getting identical deleterious recessive alleles at several genes is low. But with incest, the probability of inbreeding depression shoots up.

Consider the deleterious recessive allele that causes feeble-minded phenylketonuria when it is expressed in the homozygous state; that is, when an individual has two identical defective alleles for this gene. The defective allele exists in populations with a frequency of 0.01, meaning that if we counted the alleles of this gene from a random sample of the population, 1 out of 100 would be the defective form. Therefore, the probability of an individual being homozygous for this gene and suffering the disease is 0.01 multiplied by 0.01, or 1 in 10,000. In other words, we expect 1 in 10,000 children from outbred matings to have feeble-minded phenylketonuria. Simple laws of probability show that this risk goes way up for incestuous matings. Children from brother-sister matings are 576 times more likely to produce children suffering the disease. This is because a brother and sister share half their allelic forms. Each received half their genes from their mother and half from their father. There is a high chance—0.5, or 1 in 2—that they carry the same deleterious alleles and that their progeny will be homozygous at many genes.

Knowing the average number of lethal equivalents carried by a population enables one to calculate the mortality of children produced by primary incest. For example, if humans carry 2.2 lethal equivalents, then 42 percent of the children produced by brother-sister matings would not survive to reproduce. This is a theoretical prediction that human geneticists have documented with data from several human populations. In a study conducted in Czechoslovakia, the fate of 160 children born to women through incestuous matings with sons, brothers or fathers was compared with another set of 95 children born to the same women but with a nonrelative as the father. Close to half of the incestuously produced children failed to survive to reproduce, whereas almost 90 percent of the children of nonincestuous relationships were able to reproduce.

First-cousin marriages bear a lesser but still significant cost. In Sweden, about 4 to 6 percent of children not born of incestuous relations suffer from genetic disease. When the children come from first-cousin marriages, the incidence of genetic disease rises to 16 to 28 percent. In France, a study of children from first-cousin marriages showed that they were twice as likely to die before adulthood as children from unrelated crosses. The same difference has been reported in Japan. As expected, second-cousin marriages produce children of intermediate vigor.

The high cost of incest through inbreeding depression is well established for many other plant and animal species. We expect organisms to evolve mechanisms for avoiding inbreeding. Plants and animals can accomplish this through massive synchronized reproductive bouts or through the dispersal of one sex from the area in which it was born. Both of these may also be explained as adaptations to reduce predation or as competitive strategies. But neither predation risk nor intrasexual competition explains the ability of many animals and plants to distinguish between kin and non-kin. Some plants have recognition alleles that enable them to discriminate against pollen which is too closely related. Various marine organisms also possess recognition alleles that allow them to distinguish genetic relatives. We expect this ability in highly colonial altruistic animals such as social insects in which extended families compete with other families. But kin recognition based on odor is well developed in many mammals, including various rodents and monkeys, which have a much more flexible social system.

Humans seem to possess a similar kind of recognition system, one in which we learn to be sexually uninterested in siblings and thereby minimize incest. This argument was implied in the writings of psychologists such as Hans Westermark and Havelock Ellis, who found suggestions of a link between childhood familiarity and lack of erotic interest. The best evidence that this is so comes from studies of Israeli kibbutzim. Originally designed to be self-sufficient units, kibbutzim were composed of largely unrelated people. Children were reared communally. When the marriage patterns of a group of 2,769 children raised in kibbutzim were analyzed, it was found that not one marriage occurred between children who had been raised together. The children were largely unrelated, so the finding at first makes little sense in terms of natural selection. But communal child-rearing on this scale does not approximate the conditions under which humans evolved; for most of our history, humans have existed as nuclear families. In kibbutzim, children raised together develop the same erotic aversion to each other that exists among siblings—familiarity apparently breeds contempt.

Many anthropologists have offered different explanations for the human incest taboo, and many are openly hostile to the biological explanation based on inbreeding depression. They see the incest taboo as fulfilling a variety of strong social functions that could account for its existence. In the past, most standard anthropological explanations for incest taboos used good-of-the-group arguments, which proceed from the notion that cultural practices exist for the good of society. Freud argued that the incest taboo developed because it prevented cultural stagnation. Individuals who outbred married into social units different from the family units in which they were reared, which increased social diversity and kept the society dynamic. The same sort of argument is found in the work of such giants in anthropology as Margaret Mead and Claude Lévi-Strauss. In writing about marriage rules, Mead claims that "the primary task of any society is to keep men working together in some sort of cooperation." Lévi-Strauss concurs, saying, "The prime role of culture is to ensure the group's existence as a group," and he made famous the idea that incest taboos and marriage customs evolved as rules for wife-trading. He hypothesized that required outbreeding enables male kinship groups to trade and exchange women for material and political gain, the trading used to form political alliances that strengthen the group.

As evolutionary biologist J. Maynard-Smith has pointed out, anthropologists have made too much of the cultural significance of the incest taboo: "When Lévi-Strauss asserts that the incest taboo is the characteristic feature which originated human culture, it is in one sense tautological and in another manifestly false. If the emphasis is on the cultural connotation of the word 'taboo,' the statement is tautological, since there can be no culture without culture. If the statement implies that animals do not have practices which avoid incest, then the statement is false." He goes on: "It seems to me more plausible that the differences between the customs of different societies call for a cultural interpretation which has little to do with inclusive fitness. But to attempt a general explanation of human marriage customs and incest taboos which ignores the fact that in all probability, our ancestors avoided mating with close relatives long before they could talk would be foolishly parochial."

A more biological view of culture is that it is nothing more than an abstract construct representing the sum of individual decision-making and behavioral patterns. It is individuals who must create, implement or ignore taboos and rules and enjoy the attendant costs and benefits. Individual costs and benefits are the mechanisms that give rise to cultural patterns. It is also true that humans are social. They live in groups and compete against other groups. This creates a second and higher order of selective pressure. An individual's action is influenced not only by the immediate costs and benefits to the individual and to his or her children or kin but also by the way it will affect the fortunes of the group in which the individual must exist. Individuals may sacrifice personal gains or be coerced by other members or by historical rules to alter their selfish behavior in the interests of the group. However, it should be borne in mind that the interests of the group or the culture are genetically important only because they influence the success of the individuals and kin groups that make up the group and act out the culture of the group.

Part of the conflict between individualist and group-selection views of cultural rules is a result of the different levels of focus, the forest-and-the-trees problem. The biologist is interested in the primary-incest taboo; anthropologists have been perplexed by the complexity of systems determining marriage patterns beyond the primary-incest taboos. Many marriage possibilities that do not involve incest are prohibited. Furthermore, there is nothing in the primary-incest taboos that precludes the use of marriage rules to accomplish such ends as cementing political alliances.

Parsimony of explanation is a goal of science. But one cannot choose among factors in a multifarious process. Instead, one must build a model that incorporates them all. This is unsatisfying but necessary. To understand mating patterns, it is essential not only to factor in the cost of inbreeding depression but also to include other elements such as the costs and benefits of dispersal as well as various material considerations—the intensity of sibling rivalry, intrasexual competition and the possibility of inheriting territories and resources. The goal is not to find a single law that explains the incest taboo but to model a process and explore the operations and conflicts of various forces.

In this light, the fact that some human groups routinely practiced incest between siblings or close cousins is not a fatal flaw but material for testing a model. The people who routinely sanctioned and even encouraged incest did so for economic reasons and acted to their own personal advantage. They do not represent a random selection from human cultures. Incest was a prerogative of royalty in Iran, Peru, Hawaii and possibly Samoa. As Sir James Frazer argued in *The Golden Bough*, when the throne is passed down through the female bloodline, it profits the king to commit incest with his daughter when his wife, the queen, dies. Similarly, incest with siblings is a way to keep family inheritances and holdings intact, and the pattern of nieces marrying uncles is a way to retain dowries within the extended family. Incest is a resource-monopolization strategy when substantial material benefits are counted as more valuable than the cost of inbreeding depression. A family willing to eliminate malformed children by means of infanticide further reduces the cost of incest. This practice was well known in most incestuous groups, and the longer the incest carried on, the lower its cost, because deleterious alleles would be exposed to selection and removed from the lineage.

The costs and benefits of incest also vary for the two sexes. In general, males will experience a lower cost for an incestuous mating than will females. A woman who bears a defective offspring has substantially reduced her lifetime reproductive success, but the man who fathers it has not diminished his reproductive success much. In fact, if little parental investment is exacted from him for the offspring, he will have raised his reproductive success. Thus we expect males to be more inclined to commit incest than females. We also expect individuals nearing the end of their reproductive career to be willing to commit incest. They, too, have little to lose and much to gain.

Anecdotes prove very little, but I will recount one about the costs and benefits of incest among the Inuit as described by French anthropologist Jean Malurie in his magnificent book *The Last Kings of Thule*. "The Eskimo is above all a realist—don't worry about how you live, he says, just see that you survive—and he adapts his rules to fit the circumstances. The 'exofamilial' solution was to marry outside of the family altogether; this widened a man's choice of mate but also extended the obligations to give financial help that he incurred through his marriage. The 'endofamilial' solution for the hunter who was a widower, isolated and unable to find a second wife either among his relatives or elsewhere, was to sleep with his own daughter and so ensure himself a true partner. I know of such a case in northeastern Canada. The widower, who was a good hunter and had had many children, had spent two years in a fruitless search among neighboring groups for a new wife; only then did he decide on this course, provoking the silent disapproval of his neighbors and causing great sorrow to his oldest daughter, on whom his choice had fallen. His search had been useless, and no wives were to be had because the rate of female infanticide in the area had been very high between 1930 and 1940."

Avoidance of inbreeding depression is only part of the cost-benefit equation. It is also necessary to consider the costs of outbreeding. Dispersal itself is costly. For some insects, it can be shown that there is a trade-off between the energy consumed by dispersal and that required for reproduction. One is more likely to be eaten when dispersing, and it may be harder to acquire a territory in a new and unfamiliar area.

There are also genetic costs to outbreeding. Plants, at least, seem to become locally adapted to a certain area. The mixing of genes from distantly separated individuals will break up the match between the genotype and the environment, and it may break up particularly favorable gene sequences. In some plants, outbreeding depression takes place on an amazingly small scale. Nicolas Waser and Mary Price did the first studies documenting this effect on a small scale. They worked with *Delphinium nelsonii*, a dark blue larkspur that grows in the alpine meadows of the Colorado Rockies. They cross-pollinated plants by hand, moving pollen from plants separated by distances varying from three to more than 3,000 feet. No one was surprised that they found inbreeding depression in the crosses from only three feet apart. *Delphinium* seeds are relatively heavy and probably do not disperse far, so the plants growing within three feet of each other may be close relatives. What was surprising was that maximum seed set, an indicator of the success of the pollination, occurred in plants only 30 feet apart. The matings from a separation of more than 3,000 feet were lower.

The interpretation is that if mating partners are too dissimilar genetically, then there is a lack of compatibility in the resulting offspring. Waser and Price have shown that the natural pollinators of *Delphinium*—various bumblebees and hummingbirds—transfer little pollen beyond the optimal 30-foot distance. This means that over time, the individuals in a given spot will tend to be made up of genetic combinations that are both favorable and drawn from a highly local pool. Bringing pollen from great distances breaks up the local coadaptation of genetic combinations.

The logical extension of this argument is that plants with different seed- and pollen-dispersal systems should have different scales of optimal outcrossing. Several other works have studied optimal outcrossing neighborhoods in plants that have high rates of long-distance pollen transfer. Bat-pollinated trees or desert plants—whose seeds are washed great distances down flash-flooding arroyos—have optimal outcrossing neighborhoods in the order of several miles.

As a human example, we can consider the geographic variation of blood groups, which are genetically determined. If a woman with Rhesus-negative (Rh-negative) blood mates with a Rhesus-positive (Rh-positive) man, then she often has a so-called blue baby during later pregnancies. She develops an immune reaction to the foreign Rhesus blood-group factors found in her baby but not in her own blood. During the first pregnancy, there is usually no problem, because there is little mixing of blood between the mother and the fetus, but at birth, considerable mixing occurs. As a result, the mother develops antibodies in her blood against Rh-positive blood, and during the next pregnancy, these antibodies can enter and destroy the blood of the Rh-positive baby. Rh-positive blood is widespread in Europe, and blue babies are rare. In China, most people are Rh-negative, so there is a significantly greater risk of blue babies resulting from matings between European men and Chinese women.

Of all the things humans do, mating and reproducing would seem most likely to influence evolution. They are and always have been far more important than, say, one's taste in music. Thus it comes as a surprise that the discussion of the biology of incest avoidance should have provoked negative reactions from many scientists. Any discussion of human biology always seems accompanied by the fear that evolutionary interpretations of our behavior will justify a behavior or discourage or prevent us from changing it. For example, sociobiologist E.O. Wilson aroused the ire of many anthropologists and psychologists when he argued in *On Human Nature* that the sociobiological explanation of the incest taboo "identifies a deeper, more urgent cause, the heavy physiological cost imposed by inbreeding."

A caustic reviewer, experimental psychologist N.L. MacIntosh, counterclaimed that "biological explanations simply cannot preempt sociological explanations this way. We could accept that the biological function of incest taboos was to prevent inbreeding, but this would not exclude the possibility that they served social functions as well, and not just as secondary contributing factors. Nor would the identification of any such biological function allow one to choose between an infinity of possible proximate (social and psychological) causes." Granted, but it is also true that the social functions cannot develop uninfluenced by the genetic cost of inbreeding.

There is no doubt that a family marooned on a desert island could overcome its aversion to incest or that royalty would permit a son to bed a daughter if it meant keeping the family fortune intact or that a tribe might prohibit marriages which led to strife or weakness. Biologists concede the point that humans are able to understand, use and culturally modify biological tendencies—in this case, the avoidance of inbreeding depression. But the corollary concession is also necessary. Evolutionary history affects cultural practices. It does not rigidly bind or predict them, but it is an influence. We cannot, however we might wish it, escape the fact that we are animals and minds within bodies and within history.

CHAPTER 15

Islands of Incest

"We are too big for our world. The microscope takes us down from our proud and lonely immensity and makes us, for a time, fellow citizens with the great majority of living things. It lets us share with them the strange and beautiful world where a meter amounts to a mile and yesterday was years ago."

Asher Treat
Mites of Moths and Butterflies

Every once in a while, a nasty itch afflicts Midwesterners. In 1950, for example, the patrons of the Indiana State Fair found themselves scratching more than usual. The same itch also plagues sailors on certain types of cargo ships. Hay itch, as it is known, is a special sort of affliction accompanied by nausea, headaches, festering sores and fever. And it has absolutely nothing to do with any similarity between the hygienic practices of sailors and those of Midwesterners. It simply comes from hay-itch mites.

These mites are associated with the hay used as bedding for livestock. They are members of the genus *Pyemotes*, a name that has the delightful translation "native of pus." The female resembles other parasitic mites and ticks. After sinking her mouthparts into some animal's hide, she engorges and swells into a milky, translucent egg-filled balloon. It is the offspring of this mite that merit our attention.

Unlike most other mites, in which the immature young go through a series of developmental stages, molting as they grow larger, these mites are born as adults. All the intermediate stages have been compressed into one. The first to hatch are males. They take on the role of obstetrician, inserting their pincerlike legs into their mother and pulling out their sisters. As soon as a male gets possession of one of his sisters, he mates with her. The male probably assists his mother for selfish reasons. He is trying to speed up the process and also get first sexual access to his sisters. A male may inseminate all 20 or so of his sisters.

This is incest in the extreme, and we normally think of incest as a kind of sexual suicide. If the purpose of sex in the first place is to produce offspring with genetic diversity, why should *Pyemotes* evolve this intentional, indeed avid, and methodical congress of brothers and sisters? Why do males or females not disperse in search of less similar gametes than those of their siblings?

Every man is an island. That is what makes the intensely incestuous ways of the hay-itch mite adaptive. To a small, creeping mite, the bodies of humans and their domestic beasts can be so huge that dispersal is extremely difficult. Students of mites are finding that the world is full of hitherto unsuspected islands of time and space.

The common house sparrow is a universe as much as any man or beast. It is a far-flung archipelago for many kinds of mites. The best-studied of these mites is the quill mite, *Syringophilus minor*, a species that lives as if the tiny interior of a sparrow-feather shaft were an isolated island. Most feathers are colonized by a single female mite. She crawls inside the hollow quill, drills into the wall and begins feeding. She lays a dozen eggs of which only one, the first, is male. He, too, is a sister-mating machine. He mates with all his sisters, dies and leaves them to repeat the process. After two generations of females, each producing a dozen offspring, the shaft is packed with mites, and the females must disperse in search of new feathers. Normally, this happens when the birds are molting and growing a new set of virgin feathers or when young nestlings are fledging. At those times, thousands of untouched islands are available for colonization.

The feather shafts close quickly, however; the window of vulnerability for a feather is narrow in both time and space. For a male, the opportunity to disperse from feather to feather in search of female mates other than sisters or very close cousins is virtually nil. As for the sisters, their task is to generate as many propagules as they can to broadcast over the archipelago that will suddenly and briefly be open to colonization. This means that it will pay sisters to mate with a brother that has virtually no chance of mating with anyone else except a close cousin. Sisters may also save directly on the time spent in courtship and mating by being incestuous. It is an adaptation to the extreme insularity of their universe.

For some mites, an island is as small as an insect egg. In the Middle East, certain ornamental fig trees become infested with thrips. Thrips are small insects to begin with, being only a few millimeters long, and their eggs are smaller still. Nevertheless, a single thrips egg is enough to rear a whole family of thrips egg mites, *Adactylium* spp. Development in these mites is again a hasty, incestuous proceeding. When a gravid female mite finds a thrips egg, she fastens onto it, and as she imbibes its contents, sons and daughters begin to grow inside her. Instead of emerging as larvae, they set about eating their mother from the inside out. The male hatches first and proceeds to inseminate his sisters while still inside the womb. This is his only mission in life, and once it is fulfilled, he may die without ever being born. His sisters cut a hole through their mother's abdomen and crawl unsentimentally away in search of their own egg to suck.

The same theme is played with an Oedipal variation by the mite *Histiostoma murchei*, a parasite of the egg cocoons of earthworms in Michigan and probably other states. Earthworm cocoons are substantial structures built of a semisolid mucous envelope and packed with hundreds of eggs. Each one will feed hundreds of mites, but they are distributed willy-nilly through the soil and litter and are difficult to find. When a female mite stumbles on a cocoon, she has found a vast trove of nutrients to be exploited, the means to copy her genes to great excess. But most often, the wandering female is still a virgin. She lacks a male to provide sexual diversity in her progeny. However, she has a solution, albeit a compromise: She begins laying eggs in spite of her virginal status. She lays two to nine eggs, and all of them hatch into males within 12 to 24 days. Needless to say, these males are, by our terms, strange bedfellows.

The males remain at home with their mother. Within two days, they are sexually mature, and as you will have surmised, they copulate with their mother. Then they die. The mated mother proceeds to lay a new clutch of up to 500 eggs, all of them females. These females then disperse to follow the lifestyle of their mother. Can any mating system be more bizarre? The son is husband to his mother and father to his sister. The sister is the daughter of her brother and the wife of her son.

Mites may do even more unexpected things. In his poetic *Mites of Moths and Butterflies*, Asher Treat tells the story of ear mites that live as though the ear of a moth is an island. Female mites gain access to a moth by waiting at flowers, and when a moth visits for a sip of nectar, they scramble aboard. They begin crawling, using some unknown compass that points them toward the ear of their scaly behemoth host. They crawl across wings and legs, great plains of scale and hair, perhaps following the scent of some earlier colonist. Their final goal is the tympanum, the eardrum of the moth. Moths have two ears and two eardrums, but the mites infest and deafen only one of them. How they accomplish this is completely unknown, but it is clearly an act of prudence that is highly adaptive.

Moths are a favorite food of bats, which hunt them using high-pitched squeaks that shimmer through the air and bounce back from the objects they encounter. The bat reads the echo pattern to calculate where the moth is and snags it out of the night air with its clawed wing. Many moth species can hear the cry of the bat. This can be demonstrated on any warm summer night when moths are spiraling around a streetlamp or other light. If a cluster of keys is thrown jangling into the air or, better, if a remote control for a television set is clicked on and off, high-pitched vibrations are produced. You cannot hear them, but you can see the effect immediately. Some of the moths will dive, plummet and loop. They are taking evasive action against what they perceive as an approaching bat.

The mites, by prudently restricting themselves to the use and destruction of one ear, reduce the chances that they will end up dissolved in the gastric juices of a bat. These mites are also incestuous; males are rare, and they show no hesitation about mating with their sisters. This implies that the moth is an island with a high cost of dispersal. One island is crowded and packed with incestuous mites, while another, as near as the other side of the moth, is pure and virgin. This is the strangest part of the story, that a moth can be two islands, one thickly infested and incestuous, a beacon for new colonists, and the other seemingly identical but forbidden and uninhabited, two islands so similar but separate, flying side by side through the dark of night.

Readers may be led to suspect that incestuous mating strategies are a peculiarity of mites. Happily, this is not true. Similar strategies have been independently discovered and put to use by a variety of ants and parasitoid wasps.

Parasitoid wasps are one of nature's richest taxons. Hundreds of thousands of species exist in almost every habitat on Earth. For the most part, they feed on other insects, drilling their eggs into a host and gradually consuming it. Most parasitoids are tiny. Some fit by the

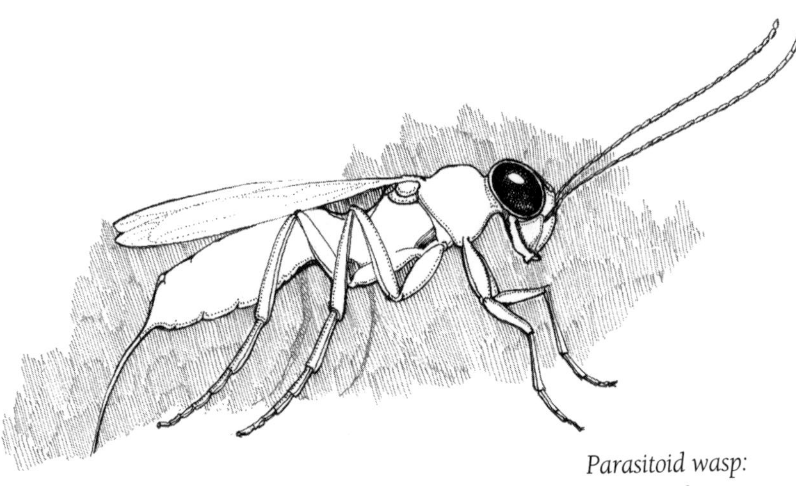

Parasitoid wasp:
The advantages of incest

dozens within a single insect egg. Thus they confront the same problem as the mites. Their hosts may be relatively huge, but they are widely and erratically dispersed in space and time. A caterpillar or a large maggot may well be an island for a parasitoid wasp. Consequently, many follow a similar pattern of incest. When a caterpillar is infested with braconid wasps, each a mere fraction of the caterpillar's bulk, it is often the clutch of a single parasitoid female that emerges. A few males emerge first, just enough to mate with all their sisters. Why produce more? The sons would only compete with each other and devalue themselves and their close kin, and their production would cost in terms of a reduced production of sisters. Sisters are the propagules sent into the wide world searching for the insular caterpillars, and colonization rates and spread of the mother's genes will be maximized by increasing the number of her propagules.

Experimental work has shown that the parasitoid *Nasonia vitripennis* is able to gauge the relative costs and benefits of dispersal, sex ratio and the chance of what is called "local mate competition" among siblings. *Nasonia* can parasitize the pupae of the blowfly. In nature, blowflies feed on a wide variety of carrion. They might feed on the carcass of a tiny shrew weighing less than an ounce or on the corpse of a cow. The former may produce but a single blowfly pupa, while the latter will yield tens of thousands. *Nasonia* males are tiny—and wimpy fliers—but they can get around among the pupae on a single carcass.

At a large carcass, where many host pupae are close together, the value of producing sons goes up. They will not compete with their brothers, and they may be able to find and inseminate the daughters of many other *Nasonia* females. At low densities, however, males will be less valuable and primarily incestuous. When experimenters provided *Nasonia* with host islands of varying sizes, the females adjusted the sex ratio of their brood accordingly. On huge islands, they produced many males, but on a single pupa, only a few sons were produced relative to the number of daughters.

Incest is adaptive not only on abstract ecological islands but on conventional ones also. If you visit oceanic islands such as those in the Caribbean and the Pacific, you'll notice immediately that social insects are rare. Even strong-flying bees and wasps are poorly represented in terms of the number of species. On the mainland, they usually depend on a mating swarm and outbreeding as part of their mating procedure. Occasional island colonists will find the procedure disrupted, so colonizations often fail.

Some ant species, however, are so successful at island-hopping and colonization that they are known in the trade as "tramp" ants. The majority of these ants possess a capacity for incest. They can breed with close relatives within or near the nest. They are free from the need for the massive synchronized nuptial flights used by so many mainland ant species, in which dozens to thousands of colonies send their sexual males and females aloft to mate in a swarm.

I had an opportunity to witness the colonizing ability of one of these tramp ants when I was a graduate student in Harvard's biological laboratories. A few years earlier, a fellow student had unwittingly imported a colony of pharaoh ants, *Monomorium pharaonis*, as stowaways in his luggage. Not a formidable-looking ant, it is tiny, pale golden, of a placid disposition and unable to sting or bite. It does, however, chew exceedingly well and is adept at discovering the tiniest crumbs of anything that may be construed as edible. And its small size enables it to tunnel along narrow channels such as the interior of electric cables and to nest in the slimmest of crannies.

In a few years, the ants had saturated the biological labs: Not a single area remained where a sandwich could be left unguarded for even a few minutes. The ants ate many an experiment. In fact, it was rumored that they had consumed radioactive isotopes, and since the building was riddled with one basic supercolony regurgitating and sharing food communally, it was suggested that the entire facility had a slight aura of low-level radioactivity. Much effort and much money were expended on trying to exterminate them with pesticides, but to no avail. The incestuous habits that made *Monomorium* a successful colonist enabled it to rebound.

Pharaoh ants mate within the nest. Again, males are rare. Only one is produced for every four or five females. This means that a single queen, able to lay both male and queen eggs, is enough to start the colonization procedure, and any single colony fragment containing a queen will be able to continue the process of reproduction, fission and colonization. Whether Harvard has been able to eliminate this pest with newer methods such as feeding them disruptive hormones, I do not know. I suspect the ants still survive, and I shudder when I reflect that there are other unoccupied warm islands so near—the many museums and libraries laden with old, precious and edible specimens.

It is probably not just coincidence that the beasts with incestuous strategies are mites and various ants and wasps. All of them share the same chromosomal method of sex determination, a system known as haplodiploidy. Most organisms are diploid; that is, they contain two sets of chromosomes in each cell, one donated by the mother, the other by the father ("-ploid" refers to the number of chromosome sets a cell or individual contains, as in haploid, diploid, triploid, tetraploid, and so on). In haplodiploid systems, males contain only one set because they are created from an unfertilized egg laid by their mother. Females are diploid, containing both a maternal and a paternal set.

This suggests two factors that may have favored incest as an adaptation to a highly insular world. We avoid incest because we are all diploid. We have two copies of every gene in our body. These two copies may be identical, or they may be of alternative forms known as alleles, forms that differ slightly in their pattern of DNA base pairs and ultimately in the molecules that their formula produces when it is translated and transcribed into proteins. One form can often be defective yet undetected; its deleterious effect—not being able to make an enzyme properly, for example—is masked by the alternative correct form. What inbreeding does is increase the chance that an individual will be made in which both copies of a gene are defective. If you have a defective allele, there is a high probability that one of your brothers or sisters or your mother or father has exactly the same defect. And if you breed incestuously, it is highly probable that your offspring will get a pair of defective alleles and suffer the consequences.

Under haplodiploidy, genetic defects are exposed every generation. The males' genes are naked in the sense that there is only one set, and any defects will be expressed. There is no second set of copies to cover deficiencies. As a result, haplodiploid genetic systems are purged of deleterious alleles in each generation. This idea is clearly supported by genetic analyses which show that regardless of their breeding system—incestuous or outbred—haplodiploids are far less genetically variable than regular diploid organisms. In other words, the genetic cost of inbreeding is lower for haplodiploids, making it easier for mites, ants and parasitoid wasps to colonize a world whose insular nature makes sexual dispersal costly.

As well, female haplodiploids have a neat mechanism for controlling the sex ratio of their progeny according to the nature of their island resources. They have an organ that controls the release of sperm, which is stored in a special sac. If producing a large number of males is appropriate, they simply withhold the sperm, and the unfertilized eggs become males. They thus have the technical means to optimize their sex ratio in a way that diploid organisms with fixed one-to-one sex ratios cannot. This makes them better suited for life in divided, locally varying populations.

The fact that creatures as unrelated as mites and various wasps and ants have converged on incest as a solution to island life is a cause for some rejoicing. Biologists are always heartened when unrelated groups evolve the same solution to similar kinds of problems. Evolutionary convergence testifies that there is pattern and sense in the solutions to the world, and it is a celebration of the predictive and unifying power of evolutionary and ecological theory if we can make sense of something as unusual as these mating systems.

But I don't believe this is the greatest importance of the incestuous life histories of mites. The wonder lies in recognizing the still unfathomed diversity and weirdness of life on this Earth, life in the soil we stand on and in the ears of moths that flutter close by us. The study of other organisms has shown us that our own behavior is no measure of the world. As rich and complex as humans are, we cover only a conservative stretch of life's possibilities. Mites and creatures like them help us appreciate the truth of a comment made by the great evolutionist J.B.S. Haldane: "The world is not only queerer than we imagine, it is queerer than we can imagine."

CHAPTER 16

Virgin Birth

"The grave hides all things beautiful and good."

Percy Bysshe Shelley
Prometheus Unbound

Entomologists usually have an affection for graveyards, especially those less-than-immaculate ones with plenty of unruly shrubbery and lichen on the tombstones. Howard Evans, the great American entomologist, points out that among the virtues of such places are the nest sites they offer for mason wasps in the incised letters on the stones. And my friend Philip Ward, himself an entomologist (of future renown), once showed me that the same spot is the ideal place to witness a particular organism's genuine and indisputably virgin birth. Her name is *Solenobia*.

Solenobia spends most of her life as a caterpillar hidden in a cryptic brown case that she drags furtively around with her. As an adult, she is wingless, and she remains chaste in her case like a nun in her cloister. There she lays her eggs and dies a virgin, never having flown, never having mated.

Hers is not a conspicuous lifestyle: On tree trunks, her more usual habitat, she is rarely noticed. It is on a tombstone that she is most easily found and appreciated. It is an existential juxtaposition. The tombstone marks the end of a unique individual, a genetic composition that will never be repeated, but *Solenobia* is a living duplicator, churning out copy after copy of herself. She reproduces parthenogenetically—that is, asexually. The eggs she lays are clones genetically identical to herself: like mother, like daughter, like mother—ad infinitum in a kind of genetic immortality.

That is not strictly true. Over time, mutations will appear, slowly altering the genome, but for the most part, *Solenobia* remains the same. This could be the most logical outcome of natural selection. If an individual is merely a gene's way of transmitting itself, what could be more direct than virgin birth? But the world is apparently too complex to accept simple copying as the way to reproduce.

Parthenogens have not inherited the earth. Of the world's some three million described species, only about a thousand depend on virgin birth. And most of the parthenogenetic species studied seem to be descended from sexual ancestors. This gives us one proximate explanation for why parthenogenesis is rare. It is difficult to transform the complicated mechanics of chromosome division, crossing over and combination that occur during sexual reproduction, as well as all the cytological (cellular) machinery, into a system that duplicates itself neatly and without difficulties. M.J.D. White, one of the world's greatest authorities on the evolutionary biology of chromosomes, calls the evolution of parthenogenesis by previously sexual organisms "a cytological tour de force" because of the technical problems involved in converting from one method to the other.

Most parthenogens appear to be recent creations that still retain traces of their sexual ancestry. Hangovers from previously sexual times can seem almost pathetic. The Amazon molly, for instance, is a parthenogenetic fish that just misses living up to her namesake, a superfemale race. These brilliant little fish, which live in the streams of Mexico and Texas, seem to get along entirely without males. Male *Poeciliopsis monacha-occidentalis* simply do not exist. Unfortunately, the eggs of the female Amazon molly require the stimulation of sperm in order to start development. The sperm chromosomes are not used by the female eggs, but the sperm is nonetheless necessary. As a result, the female molly must solicit sperm from males of another species of *Poeciliopsis*. This makes for a tricky situation. The female must rely on male indiscretion and bilk him of his sperm. From the male's viewpoint, it is an utter waste, and we must conclude either that sperm are absurdly cheap or that female mollies engage in deception. In either case, we have a clear example of how a developmental hangover in the physiology of the egg has caused the parthenogen some problems.

The same problem crops up in other parthenogens. A ptinid beetle and a salamander likewise require sperm to stimulate egg development. We see parthenogenetic dandelions, which have dispensed with sex and the need for pollination, continue to produce bright yellow flowers and nectar for insects that will pay no dividend on the flowers' sugary investment.

In the southwestern United States, parthenogenetic species of whiptail lizards, *Cnemidophorus*, have hangovers still wired into their hormonal systems. Psychobiologist David Crews has shown that all these female lizards still respond to malelike courtship. A female that is not ready to mate plays the pseudomale, which means she chases a female, grabs a fold of skin in her jaws, mounts her and imitates copulation for several minutes. This releases hormones in the female being mounted and has the effect of speeding up the rate at which she reproduces. The male act produces the same response as it did when these lizards were part of a sexual species. It is not at all clear why a female should ever play the role of a male, since she is not transmitting any genes. The only possibility is a reciprocity in which the "female" females that are stimulated and reproduce then play a male role and stimulate the previously "male" females. Some whiptail populations are composed of one giant clone in which every individual is an exact genetic copy of the others, so there may be no selective disadvantage in helping another individual reproduce. Nevertheless, the whole procedure would be better dispensed with if possible. Given such difficulties, one must wonder why any asexual species exist. What ecological force makes this secondarily derived parthenogenesis a viable alternative to sex?

The ecological feature that seems to characterize most parthenogens is weediness. Whiptail lizards are usually found in areas such as floodplains, where the vegetation is recent or frequently disturbed. One Amazonian species is apparently found only in association with humans or, more accurately, with human ecological disturbance. Similarly, parthenogenetic geckos seem to turn up in disturbed spots such as Hawaii and are immensely successful in that ultimate haven for weedy vagrants, Miami. Parthenogenetic insects have a tendency to appear in recently glaciated areas, while their sexual relics are confined to glacial refugia or more southerly areas where the glaciers never extirpated the fauna. Parthenogenetic earthworms tend to be found in rotting logs or in the leafy upper layers of the soil, while sexual species are found in the deeper, stable soil horizons. All this suggests that asexuality is favored where growth and colonization are difficult.

Cloning is not simply due to selection for colonizing ability, however. In sessile organisms, it may be favored because it facilitates the reverse, the stable occupation and domination of a habitat. The sea anemone *Anthopleura elegantissima*, which lives in the rocky intertidal area along the Pacific Coast of the United States and Canada, exists in two forms, possibly distinct species. The southern form is a sexual species that lives as solitary individuals. The more northerly variety is asexual, reproducing by dividing down the middle, and the dividends all live together. This offers members of the clone several physical advantages. Their large clusters create their own damp microhabitat that prevents harmful drying at low tide, and it may reduce the hydrodynamic drag and stress of the waves when the sea anemones are underwater. More important, cloning allows the anemone to blanket and monopolize an area of rock and exclude competing barnacles, seaweeds and worms.

Clones have a tactical advantage in group monopolization that an aggregation of sexually derived individuals can never have. All members of a clone are genetically identical, and as a result, the interests of the individual are identical to those of the group. There can be no selfishness or cheating among members of a clone. By contrast, an aggregation of sexually derived individuals generally ought to compete and to evolve cooperation only to the extent that it serves their genetic self-interest. The solidarity of clone members means they can evolve some adaptations that make their monopolization strategy more effective.

If you pull up members of several different clones and toss them all together into an aquarium, they are able to sort themselves out according to the genotype. Like settles with like. This ability to recognize genetic individuals has been studied in another colonial marine organism, the golden star tunicate *Botryllus*, which lives in mats of several hundred individuals on flat areas such as a kelp blade. Members of a clone cooperate in building an elegant device to set up water currents that run in and out of the colony bringing it a supply of food. *Botryllus* has a set of variable genes that control a series of the antigens used in immune reactions to foreign cells, the same sort of system our body uses to distinguish self from nonself. No doubt anemones have the same capability. In any case, individuals at the edge of the anemone clone have "sacrificed" their immediate reproductive interest for the competitive and defense welfare of the group. They are specialized to be "sterile warriors" that use their stinging tentacles on encroaching foreigners. Such sterility costs the individual nothing and furthers the fitness of the clone. It is exactly the same as helping itself.

Cloning to monopolize two-dimensional habitats is not only a technique of marine animals such as anemones and corals but has also evolved independently in a wide variety of plants that seek to occupy as much of a given patch of the Earth as possible. A cloning plant is concerned with the space around itself. When a strawberry reproduces, it can clone itself by sending out a runner. The runner rarely arches out for more than a foot before it begins to root and form a new plant. The distance traveled is short but relatively secure. The runner is sprouting leaves even before it roots, and the parental plant feeds the young plant nutrients through the runner to get it off to a vigorous and competitive start. A sexually generated plant is a far-flung risk. The seeds produced by sexual pollination of the flower come packaged in a juicy, bright red fruit designed for long-distance dispersal via the gut of a bird. But the seed carries only tiny resources, so its prospects of survival are slight.

One of the most glorious demonstrations of the success of cloning as a local monopolization tactic is to be seen in autumn in the Rockies when the aspens begin to color. Aspens are cloners par excellence, and when they lose their monochrome green, the various clones often turn slightly different shades of yellow. One can see a golden troop butting up to and neatly separated from a force decked out in Naples yellow. The wars of the aspen clones take place on a scale appropriate to the grandeur of their habitat. One clone of aspen has been mapped at 43 acres and is estimated to contain on the order of 50,000 trees. In the eastern part of the continent, one can see an equally beautiful battle on a less grandiose scale involving clones of staghorn sumac, which turn varying shades of crimson in autumn.

Sumacs and aspens continue to reproduce sexually as they spread clonally and, given the ubiquity of sex and the advantages of asexuality, one must ask why they bother. The answer is simply that a body is hard to build and it is a shame to waste it in death. Plants, for some reason, are better than other organisms at avoiding, or at least dealing with, the problems of old age. Plant clones constitute the world's oldest living individuals, protoplasm that has grown and remained alive for thousands of years. Aspen clones and those of the creosote bush, *Larrea tridentata*, are estimated to be more than 10,000 years old. The tenacity of the creosote bush testifies to a resilience unknown in sexual beings. In the 11,000 years since the last glacial retreat, the United States has experienced several significant climatic fluctuations. One

Aspen: Master of clones

period of aridity several thousand years ago was so intense and so long that it saw prairie plants extend as far east as Massachusetts. A relatively minor drought between about 1276 and 1299 was enough to displace much of the Anasazi culture and cause the large-scale abandonment of cliff dwellings in the southwestern deserts. But apparently it was not enough to stop the creosote bush, whose ever-growing rings still spread across the Mojave.

Another instance of the static persistence of plant clones comes from Greenland. Asexual dandelions were brought there from Iceland with Viking colonists a thousand years ago. The Viking colony in Greenland survived for 500 years, but it began to fail when the sealanes shifted and pack ice began to hem the colonists in for most of the year. The last ship known to sail from Greenland left in 1492 after a hiatus of 80 years. The isolated colonists had slowly died out. The Greenland dandelions, however, persist today unchanged, recognizable in every botanical character as identical to the Iceland dandelions.

Dandelions are often exclusively parthenogenetic, and perhaps this makes sense given their weedy nature. Yet many plant cloners such as aspens and strawberries do not rely solely on cloning. They maintain and use sexual reproduction. The evidence suggests that plants which do not soon disappear. If one looks at the taxonomic distribution of obligate parthenogenesis, it is very spotty—a few lizards here, some weevils, some scale insects, some fish in this genus, some in that family, and so on. It is a hodgepodge in which no large lineages of parthenogenetic species have evolved. Rather, habitat shifts and ecological changes bring improperly isolated species together and cause accidental hybridizations that create parthenogens. And the parthenogens thus derived never thrive for long. They do not spread and take over the range of their bisexual ancestors, and even when they persist, they do not split into other species the way sexual species do but gradually disappear instead. Thus at any single point in time, we see only scattered pockets of recently generated parthenogens.

There is no case of a parthenogenetic species replacing the sexual species from which it is derived. The reason is probably that the ability of a population to occupy and make use of ecological niches and diversity is ultimately limited by genetic diversity. Parthenogenetic species are normally composed of only a few different clones and sometimes even a single one. In this extreme case, every individual in the species is genetically identical, which is tantamount to saying that every member of the species must do exactly the same things ecologically. By contrast, a sexual species containing many genetically different individuals can occupy a range of ecological conditions. One study of this problem in the *Poeciliopsis* mollies found that the densest and presumably the most ecologically successful populations of parthenogens were those with the most genetic contact—through occasional hybridizations—with individuals from the ancestral sexual population. The contact introduces genetic variation and new clones to the population of parthenogens. Thus a multiclonal population can maintain higher numbers of individuals than a monoclonal population can. Since extinction is ultimately a question of numbers, the suggested genetic diversity may help stave off extinction.

J. Maynard-Smith, an authority on the evolution of sex, argues that the failure of parthenogens to prosper ecologically and radiate into large lineages means that we must at least consider group selection as an explanation for the persistence of sex; that is, groups of asexual organisms become extinct at a greater rate than sexual groups do. Yet the extinction of clones as opposed to sexual species is similar to selection acting on members of closely related species. The parthenogenetic individuals that clone are ultimately able to transmit and fit fewer places in the future for their asexual genes than are available for sexually generated individuals. It may take thousands of years, but asexual individuals tend to be gradually replaced by sexual ones.

As a purely sexual species, we may take some pride in being part of the obviously more successful system. Nevertheless, clones are worthy of our awe, a reminder of the brevity of our lives. The 10,000 years a clone may live is a long time, and though we may be a diverse and resilient species, as individuals we are sadly ephemeral. When I look at a patch of creosote bush or a stretching stand of aspen, I feel a wistful envy of protoplasm that may have felt the breath of mastodons or seen the first light of stars born in the time of pharaohs, soothsayers and long-forgotten prophets.

CHAPTER 17

Why Sex Persists

"Why should a coral snake need two glands of neurotoxic poison to survive, while a king snake, so similarly marked, needs none? Where is the Darwinian logic there?"

Joan Didion
Play It as It Lays

Everyone is a victim. In spite of its venomous potency, the coral snake is the evolutionary victim of the seemingly harmless king snake. It is precisely because king snakes are nonpoisonous and so similarly marked that they are a threat to the corals. Corals have evolved their distinctive red, black and yellow banding patterns as a warning to would-be predators, and they are so toxic that some snake eaters such as motmot birds have evolved an instinctive avoidance reaction to its coloration. The king snake gains protection by mimicking the coral and economizes on the price of venom and its delivery apparatus. This places the coral in an evolutionary bind. Every improvement in the fortunes of the king snake must come at the expense of the coral. When a predator learns that king snakes are palatable, it will try its fortunes on a coral, to the detriment of both. The coral has only one option: to evolve away from the king snake, perhaps achieving a new pattern, a distinctive threat, a smaller size, some reliable cue that predators can use to distinguish between the coral and its mimics. The king snake must play the role of tracker, matching the changes of its unwilling benefactor.

This coevolutionary contest is a general one; species usually achieve their fortunes at the expense of others, and they do so incessantly. The situation led Leigh Van Valen, a paleontologist and evolutionary biologist, to compare life with the predicament of Alice when she met the

Coral snake: Caveat emptor

Red Queen in Lewis Carroll's *Through the Looking-Glass and What Alice Found There*. Alice and the Red Queen run madly across a landscape resembling a giant chessboard, but when they stop, they find that nothing has moved; they are beside exactly the same tree as before.

" 'Well, in our country,' said Alice, still panting a little, 'you'd generally get to somewhere else—if you ran very fast for a long time, as we've been doing.' 'A slow sort of country!' said the Queen. 'Now, *here*, you see, it takes all the running you can do to keep in the same place.' " Evolution is like the land of the Red Queen: The pressures on an organism mean it must run—that is, evolve—as fast as it can merely to maintain itself. This, many believe, is ultimately why sex exists.

Nonbiologists often take the existence of sex for granted. It is so ubiquitous that few can conceive of life without it. Yet sex is not reproduction. There are many ways of asexually reproducing, copying oneself, without expending time and energy on mate-finding and courtship and without the increased risks of predation and contagious disease that sexual activity entails. Indeed, there are even more fundamental disadvantages to sex. For females, sex may be only half as efficient as nonsexual reproduction. One of the reasons for this is the cost of producing sons. Asexuality does away with the need for sons and allows a female to devote more energy to producing daughters.

Females are at a considerable disadvantage whenever they invest substantially more in their offspring than males do. Female eggs are normally larger and more costly than male sperm, and females often pay a higher cost for pregnancy. But both males and females get exactly the same benefits from producing an offspring sexually; each contributes half the genes. By opting for sexual reproduction, females halve the rate at which they transmit their genes. Thus females that asexually avoid sharing their maternal resources with males should do better. Apparently, they do not. In spite of its high costs, sex is virtually as old and as persistent as life itself. It is practiced by almost every kind of organism from bacteria to whales. One might think that organisms have no choice, but truly asexual species and strains within species exist. So the option is there; it is just not often taken.

There is no consensus among biologists as to why sex persists in spite of its obvious disadvantages. There is agreement on one thing, however: The undeniable virtue of sex is that it produces genetically variable offspring. G.C. Williams, the evolutionary biologist who has raised so many of the questions about why sex exists, describes selection and reproduction as a kind of lottery. The prize is genetic representation in the future, and the tickets are offspring. An asexual female can buy the most tickets because she is efficiently avoiding the costs of sons and gene-sharing. However, all her tickets have the same number: Her offspring are nothing more than copies of herself. The sexual female mixes her genes with those of another individual, and although she gets fewer tickets, they are all different. This may give a sexual female an advantage because she would be able to track environmental and biological change without relying on mutation to produce new genetic possibilities. It would enable her to run with the Red Queen.

The analogy raises two questions: Who is in the race? And how long must it be? As Williams points out, the race must be a sprint in which running is advantageous both to highly fecund species and to species such as ourselves that produce but a few offspring in a lifetime. The latter point presents difficulties for the lottery model. How can producing two lottery tickets, albeit with different numbers, be enough to compete with a strain that is producing twice as many? The other difficulty Williams foresees is one suggested by cyclical parthenogens, species such as aphids that spend part of their life cycle as sexual reproducers and another part as asexual (parthenogenetic) reproducers. If sex is to persist in these organisms, it must offer strong short-term advantages to justify the switch away from asexuality.

Enter the brain worm, *Dicrocoelium dendriticum*, not really a worm but a fluke and one that spends more of its time in innards than in brains. I knew of the complicated life of the brain worm long before I thought seriously about sex. It was only after I began to think of sex as an adaptation with costs and benefits that the brain worm's alternating sexuality became sensible. Its Byzantine life cycle is worth describing, because the way it variously employs sex and parthenogenesis suggests why sex may be valuable to all organisms, however many lottery tickets they hold.

As adults, brain worms are about an inch long, skinny and flat like all flukes and well suited for squeezing through the labyrinthine passages of the mammalian liver. Usually, they are the unwanted tenants of sheep and other grazers such as deer and groundhogs. It is their task in life to convert their hepatic home into large numbers of eggs—sexually produced eggs—which they shed into the bile ducts. Thus begins a miraculous journey, a saga that even jaded parasitologists inured to the bizarre must concede is complex beyond imagination.

The journey begins as the eggs travel down the intestines to be excreted into a pasture, where they lie quiescent until some dung-eating land snail decides to eat them. Admission to the gut of a snail has a rejuvenating effect on the dormant eggs. They awaken into roundish forms known as miricidia, drill through the gut of the snail and lodge themselves in the digestive gland. Once ensconced, they again transform, this time into an elongate form with the rather holy label of "mother sporocyte." A mother sporocyte is a cloning machine. Mothering is her only task, making vast numbers of copies of herself, known, appropriately enough, as daughter sporocytes.

As the digestive gland begins to fill, the crowded daughter sporocytes turn into yet another life stage known as cercaria. Now they look somewhat like sperm cells, with a head and a tail-like rear end that testifies to their migratory role. Their mission is to migrate to the respiratory chamber of the snail. Having hundreds of cercaria in one's respiratory chamber must be akin to having a severe cold, and the snail responds accordingly. It coats the mass of cercaria with mucus, and when it experiences a temperature drop, it expels the slime ball explosively out into the pasture.

At this point, it might seem that the snail has dealt with its parasite. It is really the other way around. The parasitic brain worm has made elegant use of the snail's physiology. The mucous coating around the cercaria has the properties of a well-designed dispersal agent. It dries into a shell that keeps the cercaria moist and alive inside and, in a general way, resembles snail eggs. The brain worm is now prepared to dupe another host.

Wood ants, *Formica* spp, have a taste for snail eggs but are not very discriminating, so when they encounter a slime ball, they lug it back to their nest for consumption. Now the brain worm begins to earn its name. Once inside the ant, the cercaria transform into metacercaria, most of which set up in the abdomen of the ant. A few, however, migrate toward the brain. Actually, they make their way to the subesophageal ganglion, a meeting of nerves that regulate much of the ant's behavior. There, they fiddle with the controls, somehow twisting the ant's mind so that it spends most of its morning and evening hours up on the end of a grass stem rigidly clamped in lockjaw. The brain worm has positioned itself for the final leg of the journey inside the infected ant, now poised to meet the bite of a grazing sheep or other animal. Down it goes to the stomach, where the sheep's pancreatic juices stimulate the metacercaria to hatch into young flukes. The flukes follow the odor trail of bile back to the liver, where they begin the sexual phase of the cycle anew.

The brain worm's insistence on sex seems at first unnecessary. Much of its task is a question of time; it is a colonist and presumably under strong selection for rapid exploitation of its host. The preeminence of asexuality as a colonization and exploitation method is not merely an argument on paper. It is known, for example, to every gardener. One week, the cabbages are pest-free, and the next, they are coated with aphids. More often than not, the outbreak on a cabbage will be the work of a single female cloning herself at top speed. If you look at one of these aphids against the light, you can see a female aphid forming within the translucent green abdomen of her mother, and sometimes, with the aid of a hand lens, even the start of a granddaughter within the daughter is visible. A real-life version of Russian dolls-within-dolls, it is a means for a mother aphid to rapidly blanket the fastness of a virgin cabbage with her descendants.

The link between accelerated propagation and asexuality is also known for other cyclical parthenogens. One of the banes of commercial mushroom farmers is a mushroom-eating midge, *Mycophila speyeri*. In the natural world, mushrooms erupt from the earth unpredictably, according to the vagaries of the weather, and are soon gone again. As an adaptation to this ephemeral resource, mushroom midges have evolved their own version of the Russian-doll strategy, but in this case, it has a formal label: paedogenesis. The word is derived from the Greek for "child" and "generation." In this strategy, the task of generation is truly an affair of the child. Child flies are, of course, maggots, and the developmental strategy of mushroom midges is to produce maggots within maggots. The colonizing fly alights on a new mushroom and lays an egg, which hatches into a maggot. Even as she eats, new maggots are forming within her, eating and bursting from their mother's belly as more maggots begin forming within them. As a result of such precocious fecundity, the mushroom soon swarms with maggots. But all sprees must end. The cabbage becomes crowded, the mushroom wasted and the sheep liver too thickly riddled with flukes. It is time for sex.

Almost inevitably, these species engage in sexual reproduction after a phase of intense asexual growth and before dispersing to a new set of hosts. There is no obvious reason why they should engage in the delays of sex. All the colonial advantages of asexual efficiency still await them. What short-term gain from sex can possibly outweigh the advantage of nonsex for these highly dispersive creatures?

For a long time, biologists thought primarily in terms of long time spans, gene pools and vast populations of individuals. From this perspective, one can argue without difficulty that sex is advantageous. It can be shown that a sexual population in competition with an asexual one will be able to accumulate and combine beneficial mutations at a vastly higher rate than a nonsexual one. By combining new genes in new combinations, the sexual population can adapt to change. But the rates involved in such projections are slow: Mutations rarely occur more often than one time in a million, and the vast majority are deleterious rather than beneficial. Normally, it requires many generations under these models to show that sexual strains can be favored over the asexual. What is needed is an advantage that can work from one generation to the next over a single season or even a weekend.

My choice of the brain worm was no accident. The explanation I favor for sex's short-term advantage is the one suggested by the Red Queen. Every species must continually adapt and counteradapt to species that would eat it, escape it or compete with it. It is significant that the brain worm becomes sexual just before it sets itself out as single eggs to be eaten by a snail. Sex will have generated a vast array of different eggs, some of which will be biochemically attuned to the variation of the snail hosts and to the changes that must be occurring in the snail population. For every generation of fluke eggs, there must be selection on the snails: Susceptible snail genotypes will be eliminated by the infection, and resistant ones will prosper. Thus the fluke (which has roughly the same generation time as the snail) must sexually generate the diversity needed to track the escaping snail genotypes.

But why, then, does it not engage in sex as it prepares to colonize the ant or the sheep? Williams' lottery analogy works here. For one thing, the brain worm colonizes these hosts in large numbers, so there is a likelihood of matching both the mammal and the ant, whose queens live for a decade or more. Both hosts have long generation times and are less able to escape the parasite genetically. The argument also works the other way. The ant and the sheep have long generation times, so a parasite with a rapid generation time (and most diseases proliferate rapidly) can adapt to its host. The host, however, in spite of its low fecundity, still benefits from sex. By producing offspring genetically different from itself, it reduces the opportunity for parent-offspring disease contagion.

This makes sense of a genetic puzzle. In the past two or three decades, new chemical techniques such as electrophoresis have revealed an unexpected amount of genetic variability in the world. Roughly 10 percent of the genes surveyed are heterozygous: An individual carries two versions of the same gene for roughly 10 percent of its genes. This came as a surprise to those steeped in a genetic theory that assumed there was such a thing as the most fit genotype and only minor amounts of variation in most populations. It is now known that some genes have as many as two dozen different forms (alleles) in a population, which led some theoretical population geneticists to conclude that much of the variation is selectively neutral, that is, it does not matter whether an individual has one form of a gene or another; the variation is merely a consequence of random forces such as mutation. This is heresy to many biologists who believe that the effects of natural selection dominate the random quirks of nature. They would like to believe that the genetic diversity of the world can be explained by natural selection. The notion of the Red Queen gives them hope, since it is supported by a growing literature on the genetics of disease.

There is now much evidence that genetic diversity, adaptation of parasite and counteradaptation of host play an important role in phenomena such as resistance to malaria, resistance of insects to insecticides and resistance of plants to insects and pathogens. A study of the struggle between flax and a rust fungus that attacks it found at least 5 genes and 26 alleles involved in resisting rust. The system is not only diverse but also dynamic. The rust is never genetically idle. A long-term study of flax-resistant genes examined and selected in 1940 and reexamined 30 years later showed that they had all become ineffective. In other words, the rust had evolved to overcome the genetically controlled defenses of the host.

Rust is only one of an abundance of pathogens that every plant must deal with. I remember being shocked when, as a result of being on the mailing list for U.S. Department of Agriculture publications, I would periodically find in my mail some massive tome devoted to, say, diseases of the potato—hundreds of commercially significant viruses, bacteria, molds, rusts and fungi. Or I might find an entire book on viruses known to trouble the rootstock of peach trees. Potatoes and peaches are probably not atypical. Everyone, every species, is the victim of a rich flora and fauna.

It is not surprising that students of the quandary of sex usually think of disease as something affecting only sick individuals. It is a problem of perception we owe to the success of modern medicine. Epidemiologist Hans Zinsser said, "Man sees it from his own prejudiced point of view; but clams, oysters, insects, fish, flowers, tobacco, potatoes, tomatoes, fruit, shrubs and trees have their own varieties of smallpox, measles, cancer or tuberculosis. Incessantly, the pitiless war goes on, without quarter or armistice—a nationalism of species against species.... The important point is that infectious disease is merely a disagreeable instance of a widely prevalent tendency of all living things to save themselves the bother of building, by their own efforts, the things they require.... The cow eats the plant. Man eats both of them; and bacteria (or investment bankers) eat the man."

As I write this, I find myself in sympathy with Zinsser's view, having just returned from a parasite checkup in Costa Rica, a very healthy country, as tropical countries go. There, the list of intestinal parasites routinely checked for includes *Ancylostoma*, *Necator*, *Ascaris lumbricoides*, *Trichuris trichiura*, *Strongyloides stercoralis*, *Enterobius vermicularis*, *Hymenolepis nana*, *Taena*, *Entamoeba histolytica*, *Entamoeba coli*, *Endolimax nana*, *Iodamoeba buetschlii*, *Giardia lamblia*, *Trichomonas hominis*, *Enteromonas hominis*, *Chilomastix mesnili* and, finally, *Balantidium coli*. And there are also a couple of spaces provided for *otros*, "others."

These pathogens are continually seeking chinks in the genetic armor of their victims, and the victims are continually trying to escape. One prediction which can be made about this state of affairs is that a sexually reproducing and short-lived pathogen will be adapting to the most common host genotypes. This is difficult to show, but a corollary can be tested. Minority host genotypes ought to be more successful than more common genotypes when they are all under pathogen pressure. This prediction has been tested using sweet vernal grass, *Anthoxanthum odoratum*, a perennial attacked by at least eight different rust species. The experimenters propagated different clones of the grass and planted them out in mixtures. Each plot contained one common majority genotype and another rarer minority genotype. They then allowed the plots to become infected. After three years, estimates of fitness showed that the minority genotypes were doing twice as well as the majority genotypes. This agrees well with an old observation of grain growers: Mixed stands are less prone to disease than large monocultures are. Similarly, it is known that the yield from a single grain plant is greater when it is surrounded by genetically unrelated individuals than when it is planted in a monoculture.

Effects of this kind ought to be even more pronounced when the generation time of the host is much different from the parasite's. The greater the difference in generation time, the more opportunity there is for the parasite to adapt to the host genetically. With this in mind, several biologists have studied the genetic interaction between pine trees and various scales that feed on them. Scales are tiny insects that may complete several generations in a summer, while a ponderosa pine may live for centuries. If there is any scope for parasite adaptation to the host, it ought to be here. And, in fact, when transplant experiments are done, the scales do best when transplanted to a clone of their original host tree. The logical conclusion is that they have become adapted to that tree, particularly since it is known that the scales have a penchant for inbreeding as they multiply on the host. Inbreeding lies between asexuality and extreme sexuality and results in less genetic diversity than matings between genetically unrelated individuals. By inbreeding, scales maintain the adaptation of their genome to that of their long-lived host.

A related set of experiments done with scales on sugar pines found that inbreeding has, as expected, the opposite consequences for hosts. Trees derived from self-pollination—the ultimate in inbreeding—had very low resistance to scales transplanted from the parent trees. Outbred trees had much higher resistance than their parent trees did. This is evidence that sex—and outbreeding sex especially—breaks up the genetic match between host and parasite. Sex reduces the chances that the pests of the parents can be visited on the progeny.

The emphasis on novelty per se is different from an earlier idea about sex's advantages. Some believed that the sexual reshuffling of genetic combinations was favored because it always produced some "elite" genotypes of particularly high fitness. G.C. Williams recognized that this argument was limited by the Sisyphean nature of the genotypes. Like Sisyphus, whose stone rolled downhill just as he approached his goal at the top, an elite genotype is destined to crumble every time sex takes place. The recombination of the chromosomes of the two partners means that new combinations of genes, different from those of either parent, will be represented in the children; sex inevitably erases the genetic aristocracy.

This leads one to conclude that there may be selection for the production of elite genotypes in highly fecund organisms such as elms and oysters, which spawn millions of descendants at a time. Selection could winnow out the chaff from every clutch, and enough of the elite would remain alive to repeat the process. In every generation and to every breeding individual, the elite would be of value. Thus the concept of elite genotypes may explain the short-term advantage of sex, but only in highly fecund individuals. In organisms such as humans, there is no scope for winnowing out the chaff. This forced Williams to conclude that sex is a historical hangover for low-fecundity mammals such as humans, a maladaptation we became locked into because we lacked the biochemical and physiological machinery necessary for asexual reproduction.

The Red Queen escapes such recourse to historical constraint. Novelty, the new variety generated by sex, is its own reward wherever and whenever individuals and species are engaged in coadapting to each other. This expectation leads to a series of predictions about large-scale patterns in the occurrence of sexuality and asexuality. In the Tropics, where biotic diversity reaches its zenith, we expect asexuality to be comparatively rare. And it seems true that parthenogenetic animals and plants are more common at high latitudes and where species diversity and pathogen pressure are low relative to biological pressures. We expect a dependence on outbreeding to be correlated with the length of generation time, and at least one survey has shown that self-fertilization is rarest in large trees, more common in shrubs and most common in herbs and wildflowers. Similarly, we expect the enzymes involved in disease resistance within the body to be more variable than those used in nondefensive tasks. Too little is known to say whether this is a general pattern, but it is known that enzymes involved in immune responses to foreign contaminants are highly variable.

But not all biologists count the need for novelty as their favorite explanation. Some believe that sexual diversity is a way for an organism to increase the number of niches its offspring can occupy or that sex helps track changes in the environment. It may reduce competition among siblings. Sexual diversity can do so many things that it is difficult to agree on a single universal theory explaining its persistence. Sex may not endure solely because of the necessity to coadapt. Yet what else but the struggle of life against life is as universal, common to brain worms and humans alike? Sex changes whatever it touches, and in that way, it has become its own ever-renewing justification.

FURTHER READING

Asterisks indicate recommended references.

PREFACE: HOW TO LOOK AT LIFE

*Alcock J. *Animal Behavior: An Evolutionary Approach*. 3rd ed. Sunderland, Massachusetts: Sinauer Associates Inc. Publishers, 1984.

*Bateson, P., ed. *Mate Choice*. Cambridge: Cambridge University Press, 1983.

Campbell, B., ed. *Sexual Selection and the Descent of Man, 1871-1971*. Chicago: Aldine Publishing Company, 1972.

*Daly, M. and M. Wilson. *Sex, Evolution and Behavior*. 2nd ed. Boston: Willard Grant, 1983.

*Darwin, C. *The Descent of Man and Selection in Relation to Sex*. 2 vols. London: John Murray, 1871.

*Dawkins, R. *The Extended Phenotype: The Genes as the Units of Selection*. Oxford: W.H. Freeman and Company, 1982.

*Emlen, S.T. and L.W. Oring. "Ecology, sexual selection and the evolution of mating systems." *Science* 197 (1977): 215-223.

*Ghiselin, M.T. *The Economy of Nature and the Evolution of Sex*. Berkeley: University of California Press, 1974.

Gould, S.J. and R.C. Lewontin. "The spandrels of San Marco and the Panglossian paradigm: a critique of the adaptationist programme." *Proceedings of the Royal Society of London B Biological Sciences* 205 (1979): 581-598.

Hapgood, F. *Why Males Exist*. New York: New American Library, 1979.

Lewontin, R.C. "Adaptation." *Scientific American* 239 (1978): 212-230.

Maynard-Smith, J. *The Evolution of Sex*. Cambridge: Cambridge University Press, 1978.

Milne, L. and M. Milne. *The Mating Instinct*. New York: New American Library Inc., 1968.

Peattie, D.C. *An Almanac for Moderns*. New York: G.P. Putnam's Sons, 1935.

Searcy, W.A. "The evolutionary effects of mate selection," *Annual Review of Ecology and Systematics* 13 (1982): 57-85.

Sebeok, T.A. *How Animals Communicate*. Bloomington, Indiana: Indiana University Press, 1977.

West-Eberhard, M.J. "Sexual selection, social competition and speciation." *The Quarterly Review of Biology* 58 (1983): 155-183.

*Williams, G.C. *Adaptation and Natural Selection*. Princeton: Princeton University Press, 1966.

_______. *Sex and Evolution*. Princeton: Princeton University Press, 1975.

Wilson, E.O. *Sociobiology*. Cambridge, Massachusetts: The Belknap Press of Harvard University Press, 1975.

Zinsser, H. *Rats, Lice and History*. New York: Bantam, 1967.

1. SPERM COMPETITION

Alexander, R.D. "On the origin and basis of the male-female phenomenon." In *Sexual Selection and Reproductive Competition in Insects*, edited by M. Blum and N. Blum, 417-440. New York: Academic Press, 1979.

*Alexander, R.D., J.L. Hoogland, R.D. Howard, K.M. Noonan and P.W. Sherman. "Sexual dimorphisms and breeding systems in pinnipeds, ungulates, primates and humans." In *Evolutionary Biology and Human Social Behavior*, edited by N.A. Chagnon and W. Irons, 402-435. North Scituate, Massachusetts: Duxbury, 1979.

Cartar, R.V. "Testis size in sandpipers: the fertilization frequency hypothesis." *Naturwissenschaften* 72 (1985): 157-158.

Cohen, J. "Gamete redundancy: wastage or selection?" In *Gamete Competition in Plants and Animals*, edited by D.L. Mulcahy. Amsterdam: North-Holland Publishing Company, 1975.

Crook, J.H. "Sexual selection, dimorphism and social organization in the primates." In *Sexual Selection and the Descent of Man, 1871-1971*, edited by B. Campbell, 231-281. Chicago: Aldine Publishing Company, 1972.

Harcourt, A.H. "Intermale competition and the reproductive behavior of the great apes." In *Reproductive Biology of the Great Apes: Comparative and Biomedical Perspectives*, edited by C.E. Graham, 301-318. New York: Academic Press, 1981.

Mane, S.D., L. Tompkins and R.C. Richmond. "Male esterase 6 catalyzes the synthesis of a sex pheromone in *Drosophila melanogaster* females." *Science* 222 (1983): 419-421.

Milton, K. "Mating patterns of woolly spider monkeys, *Brachyteles arachnoides*: implications for female choice." *Behavioral Ecology and Sociobiology* 17 (1985): 53-59.

Nakatsura, K. and D.L. Kramer. "Is sperm cheap? Limited male fertility and female choice in the lemon tetra (Pisces: Characidae)." *Science* 216 (1982): 753-754.

Parker, G.A. "Why are there so many tiny sperm? Sperm competition and the maintenance of two sexes." *Journal of Theoretical Biology* 96 (1982): 281-294.

Short, R.V. "Sexual selection in man and the great apes." In *Reproductive Biology of the Great Apes: Comparative and Biomedical Perspectives*, edited by C.E. Graham, 319-341. New York: Academic Press, 1981.

*Smith, R.L., ed. *Sperm Competition and the Evolution of Animal Mating Systems*. New York: Academic Press, 1984.

2. PENETRATING SOLUTIONS: TRANSVESTITES, RAPISTS AND DWARFS

Abele, L.G. and S. Gilchrist. "Homosexual rape and sexual selection in acanthocephalan worms." *Science* (1977): 81-83.

Monteiro, W., J.M.G. Almeida, Jr. and B.S. Dias. "Sperm sharing in *Biomphalaria* snails: a new behavioral strategy in simultaneous hermaphroditism." *Nature* 308 (1984): 727-729.

Nakatsura, K. and D.L. Kramer. "Is sperm cheap? Limited male fertility and female choice in the lemon tetra (Pisces: Characidae). *Science* 216 (1982): 753-754.

Pietsch, T.W. "Dimorphism, parasitism and sex: reproductive strategies among the deep-sea ceratioid anglerfishes." *Copeia* 4 (1976): 781-793.

Thornhill, R. and N. Wilmsen Thornhill. "Human rape: an evolutionary analysis." *Ethology and Sociobiology* 4 (1983): 137-173.

3. CONSUMING PASSIONS

Buskirk, R.E., C. Frohlich and K.G. Ross. "The natural selection of sexual cannibalism." *The American Naturalist* 123 (1984): 612-625.

Sakaluk, S.K. "Male crickets feed females to ensure complete sperm transfer." *Science* 223 (1984): 609-610.

Sillen-Tullberg, B. "Prolonged copulation: a male 'postcopulatory' strategy in a promiscuous species, *Lygaeus equestris* (Heteroptera: Lygaeidae)." *Behavioral Ecology and Sociobiology* 9 (1981): 283-289.

*Thornhill, R. and J. Alcock. *Evolution of Insect Mating Systems*. Cambridge, Massachusetts: Harvard University Press, 1983.

4. HONEST SALESMEN

Andersson, M. "Female choice selects for extreme tail length in a widow bird." *Nature* 299 (1982): 818-820.

Baker, R.R. and G.A. Parker. "The evolution of bird coloration." *Philosophical Transactions of the Royal Society of London B Biological Sciences* 287 (1979): 63-130.

Boake, C.R.B. and R.R. Capranica. "Aggressive signal in 'courtship' chirps of a gregarious cricket." *Science* 218 (1982): 580-582.

Caryl, P.G. "Telling the truth about intentions." *Journal of Theoretical Biology* 97 (1982): 679-689.

Fisher, R.A. *The Genetical Theory of Natural Selection*. Oxford: Oxford University Press, 1930.

Foster, M.S. "Odd couples in manakins." *The American Naturalist* 111 (1977): 845-853.

Fretter, V. "Prosobranchs." In *The Mollusca*, edited by A.S. Tompa, N.H. Verdonk and J.A.M. van den Biggelaar. Vol. 7, *Reproduction*: 1-45. New York: Academic Press, 1984.

Hamilton, W.D. and M. Zuk. "Heritable true fitness and bright birds: a role for parasites?" *Science* 218 (1982): 384-386.

*Kodric-Brown, A. and J.H. Brown. "Truth advertising: the kinds of traits favored by sexual selection." *The American Naturalist* 124 (1984): 309-323.

Lambert, D.M., P.D. Kingett and E. Slooten. "Intersexual selection: the problem and a discussion of the evidence." *Evolutionary Theory* 6 (1982): 67-78.

Lande, R. "Models of speciation by sexual selection on polygenic traits." *Proceedings of the National Academy of Science USA* 78 (1981): 3721-3725.

_______. "Sexual dimorphism, sexual selection and adaptation in polygenic characters." *Evolution* 34 (1980): 292-305.

Payne, R.B. "Sexual selection, lek and arena behavior and sexual size dimorphism in birds." *Ornithological Monographs* no. 33. Washington, D.C.: American Ornithologists' Union, 1984.

Rutowski, R.L. "The butterfly as an honest salesman." *Animal Behaviour* 27 (1979): 1269-1270.

Sullivan, B.K. "Sexual selection in Woodhouse's toad, *Bufo woodhousei*. II. Female choice." *Animal Behaviour* 31 (1983): 1011-1017.

Tompa, A.S. "Land snails (Stylommatophora)." In *The Mollusca*, edited by A.S. Tompa, N.H. Verdonk and J.A.M. van den Biggelaar. Vol. 7, *Reproduction*: 48-149. New York: Academic Press, 1984.

Tompa, A.S., N.H. Verdonk and J.A.M. van den Biggelaar, eds. *The Mollusca*. Vol. 7, *Reproduction*. New York: Academic Press, 1984.

5. NEW SNEAKERS

Dominey, W.J. "Female mimicry in male bluegill sunfish: a genetic polymorphism?" *Nature* 284 (1980): 546-548.

Gross, M.R. "Sneakers, satellites and parentals: polymorphic mating strategies in North American sunfishes." *Zeitschrift für Tierpsychologie* 60 (1982): 1-26.

Hogg, J.T. "Mating in bighorn sheep: multiple creative male strategies." *Science* 225 (1984): 526-528.

Perrill, S.A., H.C. Gerhardt and R.E. Daniel. "Mating strategy shifts in male green tree frogs, *Hyla cinerea*: an experimental study." *Animal Behaviour* 30 (1982): 43-48.

Potts, G.W. and R.J. Wootton, eds. *Fish Reproduction: Strategies and Tactics*. London: Academic Press, 1984.

Rubenstein, D.I. "Resource acquisition and alternative mating strategies in water striders." *American Zoologist* 24 (1984): 345-353.

Verrell, P.A. "The sexual behavior of the red-spotted newt, *Notophthalmus viridescens* (Amphibia: Urodela: Salamandridae)." *Animal Behaviour* 30 (1982): 1224-1236.

_______. The influence of the ambient sex ratio and intermale competition on the sexual behavior of the red-spotted newt, *Notophthalmus viridescens* (Amphibia: Urodela: Salamandridae)." *Behavioral Ecology and Sociobiology* 13 (1983): 307-313.

6. ROLE REVERSAL

Anderson, E. *Plants, Man and Life*. Boston: Little, Brown and Company, 1952.

*Andrews, H. et al. "Symposium: paternal behavior." *American Zoologist* 25 (1985): 779-923.

Daly, M. "Why don't male mammals lactate?" *Journal of Theoretical Biology* 78 (1979): 325-341.

Dewsbury, D.A. "Ejaculate cost and male choice." *The American Naturalist* 119 (1982): 601-610.

Erckmann, W.J. "The evolution of polyandry in shorebirds: an evaluation of hypotheses." In *Social Behavior of Female Vertebrates*, edited by S.K. Wasser, 114-168. New York: Academic Press, 1983.

Gwynne, D.T. "Sexual difference theory: Mormon crickets show role reversal in mate choice." *Science* 213 (1981): 779-780.

Hatziolos, M.E. and R.L. Caldwell. "Role reversal in courtship in the stomatopod *Pseudosquilla ciliata* (Crustacea)." *Animal Behaviour* 31 (1983): 1077-1087.

Jenni, D.A. and C. Collier. "Polyandry in the American jacana, *Jacana spinosa*." *Auk* 89 (1972): 743-789.

Johnson, Leslie K. "Sexual selection in a brentid weevil." *Evolution* 36 (1982): 251-262.

Petrie, M. "Female moorhens compete for small fat males." *Science* 220 (1983): 413-415.

Rowland, W.J. "Mate choice by male sticklebacks, *Gasterosteus aculeatus*." *Animal Behaviour* 30 (1982): 1093-1098.

Smith, R.L. "Repeated copulation and sperm precedence: paternity assurance for a male brooding water bug." *Science* 205 (1979): 1029-1031.

7. THE ECOLOGY OF ABORTION AND INFANTICIDE

Alexander, R.D. "The evolution of social behavior." *Annual Review of Ecology and Systematics* 5 (1974): 325-383.

Altmann, J. and S.A. Altmann. "Primate infant's effects on mother's future reproduction." *Science* 201 (1978): 1028-1029.

Bawa, K.S. and C.J. Webb. "Flower, fruit and seed abortion in tropical forest trees: implications for the evolution of paternal and maternal reproductive patterns. *The American Journal of Botany* 71 (1984): 739-751.

Berger, J. "Induced abortion and social factors in wild horses." *Nature* 303 (1983): 59-61.

Bruce, H.M. "A block to pregnancy in the house mouse caused by the proximity of strange males." *Journal of Reproduction and Fertility* 1 (1960): 96-103.

Day, C.S.D. and B.C. Gaff. "Pup cannibalism: one aspect of maternal behavior in golden hamsters." *Journal of Comparative and Physiological Psychology* 91 (1977): 1179-1189.

Essock-Vitale, S.M. and M.T. McGuire. "Women's lives viewed from an evolutionary perspective. I. Sexual histories, reproductive success and demographic characteristics of a random sample of American women." *Ethology and Sociobiology* 6 (1985): 137-154.

Ford, C.S. "Control of conception in cross-cultural perspective." *Annals of the New York Academy of Science* 54 (1952): 763-776.

Fuchs, S. "Optimality of parental investment: the influence of nursing on reproductive success of mother and female young house mice." *Behavioral Ecology and Sociobiology* 10 (1982): 39-51.

Goodall, J. "Infant killing and cannibalism in free-living chimpanzees." *Folia Primatologica* 28 (1977): 259-282.

*Hausfater, G. and S.B. Hrdy, eds. *Infanticide: Comparative and Evolutionary Perspectives*. New York: Aldine Publishing Company, 1984.

*Hrdy, S. *The Langurs of Abu: Female and Male Strategies of Reproduction*. Cambridge, Massachusetts: Harvard University Press, 1977.

Huck, U.W., R.L. Soltis and C.B. Coopersmith. "Infanticide in male laboratory mice: effects of social status, prior sexual experience and basis for discrimination between related and unrelated young." *Animal Behaviour* 30 (1982): 1158-1165.

Jones, J.S. and L. Partridge. "Tissue rejection: the price of sexual acceptance?" *Nature* 304 (1983): 484-485.

Labov, J.B. "Pregnancy blocking in rodents: adaptive advantages for females." *The American Naturalist* 118 (1981): 361-371.

Lee, T.D. "Patterns of fruit maturation: a gametophyte-competition hypothesis." *The American Naturalist* 123 (1984): 427-432.

Lévi-Strauss, C. *Tristes Tropiques*. New York: Atheneum, 1973.

Low, B.S. "Environmental uncertainty and the parental strategies of marsupials and placentals." *The American Naturalist* 112 (1978): 197-213.

Mallory, F.F. and R.J. Brooks. "Infanticide and other reproductive strategies in the collared lemming, *Dicrostonyx groenlandicus*." *Nature* 273 (1978): 144-146.

Mock, D.W. "Siblicidal aggression and resource monopolization in birds." *Science* 225 (1984): 731-732.

Oates, J.F. "The social life of a black-and-white colobus monkey, *Colobus guereza*." *Zeitschrift für Tierpsychologie* 45 (1977): 1-60.

O'Connor, R.J. "Brood reduction in birds: selection for fratricide, infanticide and suicide?" *Animal Behaviour* 26 (1978): 79-96.

Rohwer, S. "Parent cannibalism of offspring and egg raiding as a courtship strategy." *The American Naturalist* 112 (1978): 429-440.

Russell, E.M. "Parental investment and desertion of young in marsupials." *The American Naturalist* 119 (1982): 744-748.

Sherman, P.W. "Reproductive competition and infanticide in Belding's ground squirrels and other animals." In *Natural Selection and Social Behavior: Recent Research and New Theory*, edited by R.D. Alexander and D.W. Tinkle, 311-331. New York: Chiron Press, 1981.

Stehn, R.A. and F.J. Jannett, Jr. "Male-induced abortion in various microtine rodents." *Journal of Mammalogy* 62 (1981): 369-372.

*Stephenson, A.G. "Flower and fruit abortion: proximate causes and ultimate functions. *Annual Review of Ecology and Systematics* 12 (1981): 253-279.

Struhsaker, T.T. "Infanticide and social organization in the redtail monkey, *Cercopithecus ascanius schmidti*, in the Kibale Forest, Uganda." *Zeitschrift für Tierpsychologie* 45 (1977): 75-84.

Tait, D. "Abandonment as a reproductive tactic in grizzly bears." *The American Naturalist* 115 (1980): 800-808.

Veomett, M.J. and J.C. Daniel, Jr. "Termination of pregnancy after accelerated lactation in the rat. II. Relationship to nursing of young, day of pregnancy and length of nursing." *Journal of Reproduction and Fertility* 44 (1975): 513.

*Willson, M.F. and N. Burley. *Mate Choice in Plants*. Princeton, New Jersey: Princeton University Press, 1983.

8. FEMALE VERSUS FEMALE

Boness, D.J., S.S. Anderson and C.R. Cox. "Functions of female aggression during the pupping and mating season of gray seals, *Halichoerus grypus* (Fabricius)." *Canadian Journal of Zoology* 60 (1982): 2270-2278.

Chagnon, N.A. *Yanomamo: The Fierce People*. New York: Holt, Rinehart and Winston, 1968.

Dunbar, R.I.M. "Determinants and evolutionary consequences of dominance among female gelada baboons." *Behavioral Ecology and Sociobiology* 7 (1980): 253-265.

Dunbar, R.I.M. and M. Sharman. "Female competition for access to male affects birth rate in baboons." *Behavioral Ecology and Sociobiology* 13 (1983): 157-159.

Duslin, H.T. "Cooperation and reproductive competition among female African elephants." In *Social Behavior of Female Vertebrates*, edited by S.K. Wasser, 291-313. New York: Academic Press, 1983.

Holley, A.J.F. and P.J. Greenwood. "The myth of the mad March hare." *Nature* 309 (1984): 549-550.

Hurly, T.A. and R.J. Robertson. "Aggressive and territorial behavior in female red-winged blackbirds." *Canadian Journal of Zoology* 62 (1984): 148-153.

Koenig, W.D., R.L. Mumme and F.A. Pitelka. "Female roles in cooperatively breeding acorn woodpeckers." In *Social Behavior of Female Vertebrates*, edited by S.K. Wasser, 235-261. New York: Academic Press, 1983.

McCann, T.S. "Aggressive and maternal activities of female southern elephant seals, *Mirounga leonina*." *Animal Behaviour* 30 (1982): 268-276.

Mitani, J.C. "The behavioral regulation of monogamy in gibbons, *Hylobates muelleri.*" *Behavioral Ecology and Sociobiology* 15 (1984): 225-229.

*Small, M.F. *Female Primates: Studies by Women Primatologists.* New York: Alan R. Liss Inc., 1984.

Tiger, L. *Men in Groups.* London: Nelson, 1969.

Tiger, L. and R. Fox. *The Imperial Animal.* New York: Dell, 1971.

Valero, H. *Yanomáma: The Story of a Woman Abducted by Brazilian Indians.* London: Allen & Unwin, 1969.

Vehrencamp, S. "Relative fecundity and parental effort in communally nesting anis, *Crotophaga sulcirostris.*" *Science* 197 (1977): 403-405.

Wasser, S.K. "Reproductive competition and cooperation among female yellow baboons." In *Social Behavior of Female Vertebrates,* edited by S.K. Wasser, 350-390. New York: Academic Press, 1983.

*Wasser, S.K. and M.L. Waterhouse. "The establishment and maintenance of sex biases." In *Social Behavior of Female Vertebrates,* edited by S.K. Wasser, 19-35. New York: Academic Press, 1983.

9. MILK AND HONEY

"Breast-feeding is contraceptive." *New Scientist,* 3 May 1984: 23.

*Buss, D.M. "Human mate selection." *American Scientist* 73 (1985): 47-51.

*Cant, J.G.H. "Hypothesis for the evolution of human breasts and buttocks." *The American Naturalist* 117 (1981): 199-204.

*Frisch, R.E. "Body fat, puberty and fertility." *Biological Review of the Cambridge Philosophical Society* 59 (1984): 161-188.

Greer, G. *Sex and Destiny: The Politics of Human Fertility.* Toronto: Stoddart, 1984.

Holmberg, A.R. *Nomads of the Long Bow.* Garden City, New York: Natural History Press, 1969.

MacCormack, C.P., ed. *Ethnography of Fertility and Birth.* London: Academic Press, 1982.

*Short, R.V. "Breast-feeding." *Scientific American* 250 (1984): 35-41.

10. WHAT GOOD IS A BASTARD?

Alexander, R.D. "The evolution of social behavior." *Annual Review of Ecology and Systematics* 5 (1974): 325-383.

Bertram, B. "The social system of lions." *Scientific American* 232 (5) (1975): 54-65.

Burley, N. "The evolution of concealed ovulation." *The American Naturalist* 114 (1979): 835-858.

Galdikas, B.M.F. "Orangutan reproduction in the wild." In *Reproductive Biology of the Great Apes: Comparative and Biomedical Perspectives*, edited by C.E. Graham, 281-300. New York: Academic Press, 1981.

Knowlton, N. "Reproductive synchrony, parental investment and the evolutionary dynamics of sexual selection." *Animal Behaviour* 27 (1979): 1022-1033.

Lovejoy, C.O. "The origin of man." *Science* 211 (1981): 341-350.

Nadler, R.D., C.E. Graham, D.C. Collins and O.R. Kling. "Postpartum amenorrhea and behavior of apes." In *Reproductive Biology of the Great Apes: Comparative and Biomedical Perspectives*, edited by C.E. Graham, 69-81. New York: Academic Press, 1981.

Strassman, B.I. "Sexual selection, paternal care and concealed ovulation in humans." *Ethology and Sociobiology* 2 (1981): 31-40.

11. ORGASM AND INERTIA

Davies, E.M. and P.D. Boersma. "Why lionesses copulate with more than one male." *The American Naturalist* 123 (1984): 594-611.

*Hrdy, S.B. *The Woman That Never Evolved*. Cambridge, Massachusetts: Harvard University Press, 1981.

Kleiman, D.G. "Monogamy in mammals." *The Quarterly Review of Biology* 52 (1977): 39-69.

Morris, D. *The Naked Ape: A Zoologist's Study of the Human Animal*. New York: McGraw-Hill, 1967.

*Symons, D. *The Evolution of Human Sexuality*. New York: Oxford University Press, 1979.

12. SMELLING

Albone, E.S. *Mammalian Semiochemistry: The Investigation of Chemical Signals Between Mammals*. Chichester: John Wiley & Sons Limited, 1984.

Birch, M.C., ed. *Pheromones*. New York: American Elsevier Publishing Company Inc., 1974.

Dory, R.L., M. Ford and G. Preti. "Changes in the intensity and pleasantness of human vaginal odors during the menstrual cycle." *Science* 190 (1975): 1316-1318.

Goodwin, M., K.M. Gooding and F. Regnier. "Sex pheromone in the dog." *Science* 203 (1979): 499-561.

Gosling, L.M. "A reassessment of the function of scent marking in territories." *Zeitschrift für Tierpsychologie* 60 (1982): 89-118.

Hasler, A.D. and A.T. Scholz. *Olfactory Imprinting and Homing in Salmon: Investigations into the Mechanism of the Imprinting Process*. Berlin: Springer-Verlag, 1983.

Huck, U.W. and E.M. Banks. "Differential attraction of females to dominant males: olfactory discrimination and mating preference in the brown lemming, *Lemmus trimucrontus*." *Behavioral Ecology and Sociobiology* 11 (1982): 217-222.

_______. "Male dominance status, female choice and mating success in the brown lemming, *Lemmus trimucrontus*." *Animal Behaviour* 30 (1982): 665-675.

*Kiltie, R.A. "On the significance of menstrual synchrony in closely associated women." *The American Naturalist* 119 (1982): 414-419.

McClintock, M.K. "Social control of the ovarian cycle and the function of estrous synchrony." *American Zoologist* 21 (1981): 243-256.

_______. "The behavioral endocrinology of rodents: a functional analysis." *BioScience* 33 (1983): 573-577.

McCollough, P.A., J.W. Owen and E.I. Pollak. "Does androsterol affect emotion?" *Ethology and Sociobiology* 2 (1981): 85-88.

Muller-Schwarze, D. and M.M. Mozell, eds. *Chemical Signals in Vertebrates*. New York: Plenum Press, 1977.

*Stoddart, D.M. *The Ecology of Vertebrate Olfaction*. New York: Chapman and Hall, 1980.

Vandenbergh, J.G., ed. *Pheromones and Reproduction in Mammals*. New York: Academic Press, 1982.

13. SEX CHANGE

Borowsky, R. "Social inhibition of maturation in natural populations of *Xiphorphorus variatus* (Pisces: Poeciliidae)." *Science* 201 (1978): 933-935.

Charnov, E.L. *Theory of Sex Allocation*. Princeton, New Jersey: Princeton University Press, 1982.

Charnov, E.L. and J.J. Bull. "When is sex environmentally determined?" *Nature* 266 (1977): 828- 830.

Charnov, E.L., J. Maynard-Smith and J.J. Bull. "Why be an hermaphrodite?" *Nature* 263 (1976): 125-126.

Croll, N.A. *The Ecology of Parasites*. Cambridge, Massachusetts: Harvard University Press, 1966.

Fricke, H. and S. Fricke. "Monogamy and sex change by aggressive dominance in coral reef fish." *Nature* 266 (1977): 830-832.

Policansky, D. "Sex choice and the size-advantage model in jack-in-the-pulpit, *Arisaema triphyllum*." *Proceedings of the National Academy of Science USA* 78 (2) (1981): 1306-1308.

*_______. "Sex change in plants and animals." *Annual Review of Ecology and Systematics* 13 (1982): 471-495.

_______. "Size, age and demography of metamorphosis and sexual maturation in fishes." *American Zoologist* 23 (1983): 57-63.

Robertson, D.R. "Social control of sex reversal in a coral reef fish." *Science* 177 (1972): 1007-1009.

Shapiro, D.Y. "Serial female sex changes after simultaneous removal of males from social groups of a coral reef fish." *Science* 209 (1980): 1136-1137.

Warner, R.R. "Mating systems, sex change and sexual demography in the rainbow wrasse, *Thalassoma lucasanum*." *Copeia* 3 (1982): 653-661.

_______. "Mating behavior and hermaphroditism in coral reef fishes." *American Scientist* 72 (1984): 128-136.

14. INCEST AND OUTCEST

Alstad, D.N. and G.F. Edmunds, Jr. "Selection, outbreeding depression and the sex ratio of scale insects." *Science* 220 (1983): 93-94.

Bateson, P. "Sexual imprinting and optimal outbreeding." *Nature* 273 (1978): 659-660.

Cavalli-Sforza, L. and W.F. Bodmer. *The Genetics of Human Races*. San Francisco: W.H. Freeman and Company, 1971.

Haigh, G.R. "Effects of inbreeding and social factors on the reproduction of young female *Peromyscus maniculatus bairdii*." *Journal of Mammalogy* 64 (1983): 48-54.

Halpin, Z.T. "The role of individual recognition by odors in the social interactions of the Mongolian gerbil, *Meriones unguiculatus*." *Behaviour* 58 (1976): 117-130.

Hoogland, J.L. "Prairie dogs avoid extreme inbreeding." *Science* 215 (1982): 1639-1641.

Lévi-Strauss, C. *The Elementary Structures of Kinship*. Boston: Beacon, 1969.

Maynard-Smith, J. *The Evolution of Sex*. Cambridge: Cambridge University Press, 1978.

Shields, W.M. *Philopatry, Inbreeding and the Evolution of Sex*. Albany, New York: State University of New York Press, 1982.

Waser, N.M. and M.V. Price. "Pollinator behavior and natural selection for flower color in *Delphinium nelsonii*." *Nature* 302 (1983): 422-424.

Wu, H.M.H., W.G. Holmes, S.R. Medina and G.P. Sackett. "Kin preferences in infant *Macaca nemestrina*." *Nature* 285 (1980): 225-227.

15. ISLANDS OF INCEST

Elbadry, E.A. and M.S.F. Tawfik. "Life cycle of the mite *Adactylium* spp. (Acarina: Pyemotidae), a predator of thrips eggs in the United Arab Republic." *Annals of the Entomological Society of America* 59 (1966): 458-461.

Kethley, J. "Population regulation in quill mites (Acarina: Syringophilidae)." *Ecology* 52 (1971): 1112-1118.

*May, R.M. "When to be incestuous." *Nature* 279 (1979): 192-194.

Treat, A.E. *Mites of Moths and Butterflies*. Ithaca, New York: Comstock Publishing Associates, 1975.

16. VIRGIN BIRTH

Birky, C.W., Jr. "Parthenogenesis in rotifers: the control of sexual and asexual reproduction." *American Zoologist* 11 (1971): 245-266.

Cole, C.J. "Evolution of parthenogenetic species of reptiles." In *Intersexuality in the Animal Kingdom*, edited by R. Reinboth, 340-355. Berlin: Springer-Verlag, 1975.

Cook, R.E. "Clonal plant populations." *American Scientist* 71 (1983): 244-253.

*Cuellar, O. "Animal parthenogenesis: a new evolutionary-ecological model is needed." *Science* 197 (1977): 837-843.

Francis, L. "Contrast between solitary and clonal life styles in the sea anemone, *Anthopleura elegantissima*." *American Zoologist* 19 (1979): 669-681.

Harshman, L.G. and D.J. Futuyman. "Variation in population sex ratio and mating success of asexual lineages of *Alsophila pometaria* (Lepidoptera: Geometridae)." *Annals of the Entomological Society of America* 78 (1985): 456-458.

Herbert, P.D.N. "Obligate asexuality in *Daphnia*." *The American Naturalist* 117 (1981): 784-789.

Jaenike, J. and R.K. Selander. "Evolution and ecology of parthenogenesis in earthworms." *American Zoologist* 19 (1979): 729-737.

Vasek, F.C. "Creosote bush: long-lived clones in the Mojave Desert." *The American Journal of Botany* 67 (1980): 246-255.

White, M.J.D. *Animal Cytology and Evolution*. 3rd ed. Cambridge: Cambridge University Press, 1973.

17. WHY SEX PERSISTS

Bell, G. *The Masterpiece of Nature: The Evolution and Genetics of Sexuality*. London: Croom Helm, 1982.

Bernstein, M. "Recombinational repair may be an important function of sexual reproduction." *BioScience* 33 (1983): 326-331.

Carroll, L. *Through the Looking Glass and What Alice Found There*. Toronto: Macmillan, 1968.

Cherfas, J. "When is a tree more than a tree?" *New Scientist*, 20 June 1985: 42-45.

Dogiel, V.A. *General Parasitology*. 3rd ed. London: Oliver and Boyd, 1964.

Ferrari, D.C. and P.D.N. Herbert. "The induction of sexual reproduction in *Daphnia magna*: genetic differences between Arctic and temperate populations. *Canadian Journal of Zoology* 60 (1982): 2143-2148.

Flor, H.H. "Current status of the gene-for-gene concept." *Annual Review of Phytopathology* 9 (1971): 275-296.

Graham, J.B. and C.A. Istock. "Gene exchange and natural selection cause *Bacillus subtilis* to evolve in soil culture." *Science* 204 (1979): 637-639.

Levin, D.A. "Pest pressure and recombination systems in plants." *The American Naturalist* 109 (1975): 437-451.

Maugh, T.H. II. "Accounting for sexual reproduction." *Science* 202 (1978): 1272-1273.

Rice, W.R. "Sexual reproduction: an adaptation reducing parent-offspring contagion." *Evolution* 37 (1983): 1317-1320.

Rose, M. and F. Doolittle. "Parasitic DNA: the origin of species and sex." *New Scientist*, 16 June 1983: 787-789.

Rose, M.R. "The contagion mechanism for the origin of sex." *Journal of Theoretical Biology* 101 (1983): 137-146.

INDEX